AF577316

Physico–Chemical and Microbiological Characters of Water

Physico–Chemical and Microbiological Characters of Water

Dr. Manish L. Srivastava

2002

DAYA PUBLISHING HOUSE
Delhi - 1100 035

ISBN 81-7035-267-3

Published by : **Daya Publishing House**
1123/74, Deva Ram Park
Tri Nagar, Delhi - 110 035
Phone : 7103999
Fax : (011) 7199029
e-mail : dayabooks@vsnl.com
website : www.dayabooks.com

Showroom : 4762-63/23, Ansari Road, Darya Ganj,
New Delhi - 110002
Phone: 3245578, 3244987

Laser Typesetting : **Classic Computer Services**
Delhi - 110 035

Printed at : **Chawla Offset Printers**
Delhi - 110 052

PRINTED IN INDIA

Dedicated to

Mother Ganga

Preface

This book has been written in an attempt to explain the use of micro-organisms in removing pollutants from the environment, and the ways in which they can be utilized to recycle essential materials.

However, microbes are everywhere. In the air, in soil, in water, on our skin and hair, in our mouths and intestines, on and in the food we eat. They make the soil fertile, they clear up the environment, improve our food, they make vitamins for us inside ourselves, some protect us from less desirable microbes.

The present collection of papers, articles, international consequence of industrial pollution. Atmospheric marine and biospheric pollution have attained serious proportions, due to the lack of adequate safeguards by the polluting nations to foresee long-term consequences. The recent conflict in the Uttaranchal region is an eye-opener in this respect.

I thank Mr. Anil Mittal of Daya Publishing House for his cooperation in production of this book. Finally, I thank Mr. V.D. Upadhayay (Holland), Mrs. Samita Srivastava (Mexico), Dr. Smith (U.S.A.), Mr. Ankur Panda (Singapore), Dr. Monica Sinha (Germany), Dr. M. Parent (Japan), Dr. Jeannette Navile (Canada) and Dr. Mansi Saxena (South Africa) their help in numerous ways of this book

Dr. M.L. Srivastava

Contents

Chapter 1
Different Industrial Effluents

Introduction

Water is one of the abundantly available substance in nature. It forms about 75% of the matter of earth's crust. It is an essential ingredient of animal and plant life. About 22 million hectare metres of ground water are required for industrial processes.

Pollution of land, water and air through water generated as a result of increasing population, Urbanization and industrialisation is a challenge of serious dimensions.

According to Shende and Sundersan (1978), about 13163 gallon of water is released every day from various urban and rural areas.

The pollution of streams by industrial wastes has its beginning at the time of industrial revolution during later half of the eighteenth century. Most of the people of that era however were acustomed to a lack of sanitation and to unbeautiful surroundings. Environment disturbed by discoloured, Malodorous and healthful streams. It will be surprising if there were not some complaints or informal grumblings. The water streams become grossly polluted. The killing of fish in water or disappearance of fish from stream or chemical taste of the fish which remained and this is due to textile mills, chemical plants, coke and gas plants, Kreweries and other industrial establishments. Moderate quantities of these wastes can be absorbed without damage to a healthy stream but when these wastes are discharged in excess the water become grossly polluted.

Pollution of natural water by industrial waste is objectionable and damaging for many varied reasons. Primary importance is the possible hazard of public health by the contamination of stream to with disease producing bacteria. Another effect is of heavy metal, acids, radioactive metal and flammable liquids cause a serious problem. Sometime sewer explosions caused by gaseous vapours, methane and other flammable gases. Industrial effluents also destroy the recreational use of water. Economic damages attributed to industrial waste pollution are evident in many forms and may be considerable in amount. Damages to riparian properties is evident in corrosion of steel and concrete structure and produce nauseous gases as H_2S, also blackens the structures near the shores. Other riparian damage is loss of aesthetic value drastic decline in real estate value. Industrial waste pollution also interferes with navigation by causing corrosion of ships, docks by forming sludge bank on river bottom.

Tripathi and Pandey (1995), presents that industrial pollution reached to a alarming situation. Less than 5% industries have adopted the adequate measures for the treatment of effluents and even most of them have neglected this totally. Most of the factory wastes include the following things :

(a) Oil : forming a thin dispersed film and reduce the oxygen up.

(b) Detergents : reduce oxygen absorption capacity.

(c) Heavy metals : cause the major effect on internal body activity.

(d) Poisonous chemicals : as sulphite and sulphates.

(e) Defective micro–organism : from the research laboratories and pharmacy.

At present fertilizer involved as greatest pollutants in water stream. Metropolitan cities are facing serious pollution problem due to rapid industrialisation. The Kallu river near Kalyan affords highly acidic water (1.5 pH) due to industrial discharge. National Environmental Engineering Research Institute (NEERI) concluded that Hoogly river near Kolkata, Ganga from Kanpur to Allahabad, Damodar near Durgapur, Yamuna near Delhi and Agra, Gomati near Lucknow flow under similar state of pollution. In last years due to extensively polluted Yamuna river world's biggest epidemic infective Hepatitis is reported in which thousands of persons were died. Similar pollution condition cause the biggest epidemic of 'Jaundice' in Pimpri township and 'typhoid fever' in Sangli.

All India Institute of Public Health and Hygiene, Kolkata and Indian Council of Medical Research, New Delhi were first to take attention of problem of water pollution. NEERI, Nagpur and other institutes also started to do water pollution work.

Surface water often have taste and odour problem, particularly in region of dense population. These are generally caused by industrial and domestic pollution, algae, aquatic vegination, decaying vegetation and run off from agricultural fields (Baker, 1961). Anobena Anacystis, cosmarium, Fragillaria, chlamydomonas etc. make water unhygenic and produced undesirable things in the impounded water (Symons, 1956). Venkateswarlu in (1981), reported that growth of different algae was influenced by different factors. Several algae reduces the water quality by producing massive growth. Certain algae grow in water polluted with organic wastes and play an important role in self–purification of water. Algae also acts as indicator of water pollution in various ways. The application of biological data in assessing other water quality was firstly reported by Kolkwitz and Marsson (1908) in their "saprobein system".

For the quantifying the biological data an another system is reported by Beak (1964), Palmer (1969) and Patric (1949) use algae as indicator of water pollution. An another system for biological assessment of water pollution was produced by Patric (1953) and Wurtz (1955).

Zagic (1971) classified the water pollution courses under six categories.

(a) Poisonous, (b) suspended solids, (c) organic matter, (d) non–poisonous salt, (e) simple deoxy generators, (f) Heat.

It is stated that Kanpur is a most thickly polluted and populated city known as "Manchester of India" in density and intensity. The degree of pollution depends upon many factors including, population and type of industry and amount of water of stream in which waste is dumped. A major factor of pollution by industrial waste is discharged of organic matter. Other pollutional characteristics of industrial waste is dissolved matter, oil, grease, colour, taste, odour, acidity and alkalinity, pH and temperature. Some more important industries from the waste view point are listed below :

- Producing oxygen consuming waste (High BOD) : Beet sugar refineries, Breweries, Dairies, and Textile mills.
- Producing waste with high suspended solids : Breweries, coal washeries, distilleries, Tanneries.
- Producing oily and greasy waste : Metal finishing shops, Petroleum refineries, oil fields, Tanneries.
- Producing toxic wastes : Chemical plants, Atomic plants, Pulp mills, Tanneries.
- Producing acid wastes : Coal mines, Iron and steel mills, Chemical plants.
- Producing alkali wastes (high pH) : Chemical plants, Tanneries and Textile mills.

All the wet processing industries are included in high or liquid pollution industries. Solid waste resulting from non–processing industry as wood working and assembly operations factories. These wastes are salvaged or reduced in volume by incineration. Nuisance gases and dusts are inevitable in many industries. These are dispensed in to atmosphere unless special treatment plant is provided as electroplating operations pulp mills liquors.

Many treatment processes are used in the industrial waste treatment have been adopted from sewage treatment methods. Some time without change or some time with great or small modifications. Typical of these procedures are sedimentation, chemical coagulation, biological oxidation and chlorination with activated sludge or trickling filter. Only after a higher degree of treatment this may be suitable for recreational and irrigation purpose.

Industrial waste water also a cause of solids contamination of soils. Sludge formed in processes destroy vegetation and result in formation of lifeless areas on the earth surface. Patric (1950) has observed that pollution reduces the number of spp. and destroy the balance of life in stream. Pollution is evidenced by the biological indices of community diversity. According to Chaudhary (1979) chairman of CBPCWP the industries of Bombay account for 13% of the total waste dumped in to water bodies.

Salim (1983) concluded that heavy metal found in natural water in various states such as free ions, complex ions, and absorbed on suspended particles. The state of ion is very important factor to study the toxicity of metal. In natural water more ions are complexed than are absorbed. However the role of absorption has often been under investigated. Absorption on to suspended particle has been studied as a mode of transportation of metals.

Bernard (1987) reported a automatic detection of coliform bacteria to ensuring the industrial waste from the view of fecal contamination and tele transmission technique could be used.

Kolev *et al.* (1996) stated antibiotics are extracted from their culture media by means of butyl acetate or butanol and both solvents are removed from an antibiotic industry effluents with water azeotrophic mixture as byeproduct.

In the present work an attempt has been made to estimate the effluent quality of different industries BHEL (Bharat Heavy Electricals Ltd.), Sugar Industry and IDPL (Indian Drugs and Pharmaceuticals Ltd.).

IDPL Antibiotics (Veerbadhra, Rishikesh) : Situated 6 km. from city on Rishikesh–Haridwar Road. It is a Antibiotics producing industry. It has good quality Treatment plant [Biological Treatment Plant (BTP)], treated effluent directly dumped into Ganga by a big channel. The effluent has main impurities of organic load, (media components), heavy metals, least amount of Antibiotics, fecal contamination is more.

BHEL (Ranipur, Haridwar) : This industry is situated near Ranipur away from Haridwar city about 8 km. It is a metal industry product is Turbine, electrical tools etc. BHEL has its own treatment plant but the treated effluent goes to sewer line by a big channel and ultimately goes to Jagdeetpur sewage treatment plant. Effluent has main impurities of grease and oil, heavy metal and fecal contamination.

Sugar Industry (Laksar, Haridwar) : Industry situated 2 km. from Laksar city. Treated effluent is drain out in to a field by a long channel and ultimately dumped in to river. Effluent has main impurities of solids, fecal contamination and saccharides.

Review of Literature

The pollution of streams by industrial wastes must have its beginnings at the time of industrial revolution in later half of the eighteenth century.

The nature of industrial wastes depends upon the industrial processes from which they originate. Industrial waste water vary in nature, relatively clean rinse water to waste liquors that are heavily laden with organic or mineral water or with corrosive, poisonous, inflammable or explosive substances. Some industrial waste adhere to sewer and clog them. Acid and hydrogen sulphide destroy cement concrete and metals, hot wastes crack tile and concrete. Poisonous chemicals quite apart from their immediate danger to man but interfere with biological treatment processes and kill the organisms that normally populate in receiving water.

The major pollution problem from industrial waste lies in the disposal of organic wastes. The major source of organic waste are the food processing, paper, textile, petroleum and chemical industry.

Pollution from chemical wastes are evolved from plants manufacturing acids, detergents, explosives, insecticides, fungicides, plastics etc. Distillation, filtering and screening also produce chemical wastes.

The presence of small amount (0.2 ppm) of free ammonia in natural water is only an indication of recent organic pollution, and in highly saline water ammonia concentration ranges upto 0.05 ppm only (Hood, 1966). First fish mortality was observed in the ponds of Reasi and in Nehru stream in Jammu and Kashmir where D.D.T. was used in city drains (ICAR, 1967). Edward (1977) observed the industrial effluent are perhaps the most important sources of contamination for rivers and estuaries, whereas large scale spraying and run off water are important contributes of pesticides pollution in ponds and lakes.

Siegal and Eshleman (1975), heavy metals are the leading source of the contamination of the aquatic environment. Household bleaching solutions contain 17.24 lb of mercury and NaOH 53–1290 ppb. Malanchuck and Gruending (1973), studied the relative susceptibility of lead to the five spp. of phytoplanktons chlorophyta, chrysophyta.

Shastry *et al.* (1972), reported that waste water discharged into the Chambal river and zinc concentration ranging from 1316–6689 /1.

Robert *et al.* (1974), stated that soluble organic matter contribute to the eutrophication as to dissolved oxygen (DO). Depletion in receiving water, they have found the average pH, total solids, COD, BOD, where 7.14, 358 mg/l, 412 mg/l and 260 mg/l respectively and concluded that septic tank treatment of waste water may not be satisfactory for environmental pollution control.

CBPCWP (1985–86), pointed that the organic carbon load in terms of BOD and COD of raw and partially treated sewage and untreated industrial waste water has shown that the soil can absorb this load without much difficulty. Tambe (1986), claims 99.9% of BOD removal when treatment scheme includes anaerobic digestion followed by trickling filter and activated sludge unit.

Sharma *et al.* (1983), describes a new specific calorimetric spot test for the detection of malathion in residues water. Activated charcol is used for recovery lower limit of detection in 1 mg/l^{-1}.

Salim (1983), studied the effect of chemical composition and particle size of suspended particle in river water on the absorption of lead on to these particle and lead has been determined using the anodic voltametry. While physical and chemical parameters were used in monitoring water pollution in general during recent years, attempts towards evaluating biological methods by band of scientists have been made and results are quite encouraging. The application of biological data in assessing the quality of water was first reported by Marsson (1908) in their 'Saprobein system'.

Duzzin and Pavoni (1988), stated that heavy metal may occur in streams as a result of natural and anthropogenic factors. Concentration of pollutant were found to be related to sediments to be good pollution indicator as they accumulate all metals.

McKay et al., (1988), studied the absorption of four pollutants in aquous solution on to activated carbon, the solutes are phenol, P. cholrophenol, sodium dodecyl sulphate, and mercuric ions. *Singh (1975)*, studied the fertilizer tolerance of blue green algae and their effect on heterocyst differentiation.

Millington *et al.*, (1988), investigated the effect of varying growth medium components on the toxicity of four chemicals to three spp. of fresh water green algae by the standard algal growth inhibition test.

Genjatulin (1990), study performed which has been based on mathematical method of test culture response to action by a number of chemical water pollutants. Joy (1989) and Balakrishnan *et al.* (1990)

study the ecology of phytoplankton production in the river Periyar which receives continuous effluent discharge from a dozen industrial effluent units industrial zone of river was observed to have a high standing stock of phytoplankton during premonsoon (summer). Litreature on characteristics of industrial and sewage water and its effect on algal flora is scarce.

Kumar *et al.* (1974), carried out ecological studies on algae isolated from effluents of oil–refinery, a fertilizer and a brewery. Khare and Sharma (1979) studied the ecology of solah sugar pond at Ujjain. Palmer *(1969)* in his valuable review on algae as biological indicators of pollution found certain algae tolerant to relatively organic waste.

Ghazali *et al.* (1988) studied the six sites along the Al–Khair river in Baghdad were allocated for monitoring water pollution, escherichia coli (E.coli) was used as a fecal pollution indicator. Strain of this bacteria were tested for their resistance to many antibiotics, multiplicity of resistance and for the transferability of antibiotics resistance determinants. This study demonstrated that the Al–Khair river heavily contaminated with the antibiotic resistance E.coli which prove a capability of spreading antibiotics resistance.

Osborne and Davis (1987) collected physico–chemical and biological samples from 12 sampling stations over a 13–month period to assess the effect of a small town's chlorinated sewage and a thermal discharge on the sheep river's macro–invertebrate communities. Jolley *et al.* (1975–85) stated environmental and health concerns associated with chlorination of municipal sewage and industrial discharges were the imputes behind several international water chlorination conferences.

IAWPRC (1985) held a international Singapore conference between (28–31 May) together with (ESS) Engineering Society of Singapore and with (ASEAN) Association of SE Asian Nations. Conference on Industrial Water Technology, Treatment, Re–use and Re–cycling. All are extremely interested industrial development and concerned about the potentially adverse impact of this on the environment. The conference will deal with the range of industries most appropriate to the ASEAN Region. Discussion will centre on optimum techniques for water treatment and re–use.

Eloranta (1983) studied the effect of cooling water discharge from one small thermal power plant on physical, chemical, biological properties of water in a natural closed pond and principal changes in surface water lager were increased phytoplankton growth was evident in February causing a rapid decrease in silica, phosphate and total phosphorous concentration. Davis (1948) carried out some studies on effect of industrial waste pollution in the lower communities with special reference to effect of copper. Kumar (1965) studied the effect of certain toxic chemicals and mutagens on the growth of the blue green algae.

Rao and Rani (1985) studied the effect of Mercuric chloride on photosynthesis and respiration on biopotential. Lackey (1942) investigated the effect of distillery wastes on microscopic flora and fauna.

Kovacs *et al.* (1992) done a survey of the biological and chemical characteristic of effluents without secondary treatment from Canadian newsprint mills. The impact of these effluents on photo bacterium (microtox test) the growth of algae gives less sensitive response than even acute lethal toxicity. Webb (1985) describes the cause and prevention of sewage fungus growth in rivers receiving paper and board mills effluents.

Abeliovich (1985) work has been done to develop a biological treatment for chemical industry effluent. A multistage chemostat photosynthetic operated continuously by means of stabilization ponds. These removes the 70–80% DO, 80–85% BOD with the pesticidal activity.

Riedel *et al.* (1990) studied a microbial amperometric sensor for the determination of biological oxygen demand BOD using Trichosporon cutaneum cells immobilized in polyvinyl alcohol has been developed. This sensor allows BOD measurement with very short response. The sensitivity and specificity was increased by incubation of the BOD sensor.

Ubom and Tsuchiay (1988) investigate a simple, fast and sensitive method comprising in on chromatography, eluent suppression columns and spectrophotometric method was used to determine the iodine concentration in water sample.

Rajczyk (1993) treated the waste water formed during production of citric acid and foder yeast and carried out under dynamic condition in anaerobic up flow biofilter and methane fermentation of waste water was carried out and removal of 73%

BOD was achieved in a biofilter. Borrego *et al.*, (1990) capability of coliphages as indicator of fecal pollution was tested on basis of their survival and infectivity. The results obtained indicated that coliphages shown an similar inactivation rate to salmonella typhi and phage productive infection of E.coli cells depends upon the physiology of host bacteria. Kim and Anderson (1990) studied the effect of biological treatment on COD absorption of waste water containing eight metal cutting fluids. It is revealed that absorption capacity activated carbon increased substantially after the waste water was biologically treated. Jusiak *et al.* (1984) attempt was made to use an algal rotating disk for biological purification of nitrogen fertilizer industry effluents. Removal of ammoneim by stichococcus bacillaris growing on rotating disk followed by denitrification bed which removed about 90% of nitrogen from waste water and appearance of NO_3 and NO_2 in the purified waste water caused by the activity of nitrifying bacteria. To ensuring the sanitary quality of potable water supply requires continuous control. Traditionally germs indicating fecal contamination are detected as it is difficult to isolate or enumerate pathogenic organisms. Bernard *et al.* (1987) describes the steps we have taken to manufacture an on–site locatable automatic device (E. colimeter) for detection of coliform bacterium with response time below 10 hours.

Srivastava and Jain (1985) developed a solid membrane electrode selective to sulphate ions has been prepared from hydrous thorium oxide gel with polystyrene as binder for estimation of sulphate ions from industrial waste.

Hynning (1996) developed a procedure for the separation, identification, quantification of industrial effluents. Effluents were first extracted with a mixture of hexane and t–butyl methyl ether and then fractionated into groups. This method validated using a spiked sample and an effluent.

Garrote *et al.* (1995) used a co–agulation/flocculation method for treatment of tannery effluent on which alkaline $FeCl_3$ is used as flocculating agent and $CaOH_2$ as base/precipitant. This method reduces the chemical oxygen demand COD by 87% producing a colourless, odorless waste water. Schineder *et al.* (1995) treat the soyabean protein bearing effluent by dissolve air foltation and by sedimentation.

Kolev and Semkov (1996) studied that waste water from the antibiotics production containing quantities of butyl acetate and butanol installation, heat exchange and stripping column are used and investigation shows a degree of purification of 99–100% at different stages of operation.

Laxen *et al.* (1983) presented a scheme for the speciation of metals applied to the Pb; Cd and Cu. In the treated effluent of a lead acid battery the result are related to the physical and chemical interactions.

Suess (1982) gives a book review on standardized method of water examination and sampling. Data analysis and laboratory equipment, ASTM (American Society of Testing and Materials) which provides the world's largest source of voluntary standards.

Horne and Bennison (1987) produced a design for a three channel, recirculating laboratory stream system, suitable for use in biological experimentation. Water flow temperature, light, substrate and water chemistry can be controlled individually as desired for logic condition.

Olach *et al.* (1988) study was performed to evaluate the influence of β–cyclodextrin (β–CD) complexation on the toxicity of some pesticides (Karathane, IMID etc.) and pesticides decomposition products in activated sludge system. For a pharmaceutical waste β–CD shows a higher tolerance to pesticides.

Jenke and Frank (1985) modified the computer programme R–EDEQL–FPAK to allow for the production and simulation of chemical effect of mining aquous solution. The programme is also used to predict the optimum treatment effluent composition and provide information which defined the mechanism controlling the treatment process.

Byrne *et al.* (1988) gives an achievement to an adequate dilution. The rate of discharges effluent has been restricted on the basis of calculation made using a formula adopted by International Maritine Organisation (IMO formerly IMCO). Measurement of waste concentration in dumping vessel were made on 3 days under different climatic conditions.

Material and Methods

(A) Sampling Sites
(B) Sampling Methods
(C) Water Analysis Methods

(A) Sampling Sites

Effluent samples were collected from different industries listed below:

	Name of Industry	Manufacturing product	Treatment method
1.	RBNS Sugar Mill, Laksar	Sugar	Through biological treatment process
2.	BHEL Ranipur, Haridwar	Heavy Electrical Machines	Treated effluents goes to sewer station
3.	IDPL Antibiotic, Rishikesh	Different Antibiotics	Through biological treatment process

(B) Sampling Methods

(a) Cleaning and Sterilization of glass wares:

Chromic acid was used for cleaning the glass wares.

(i) Sampling Bottles:

300 ml capacity BOD bottles made of Borosil ware used, they were washed with chromic acid and rinse with tap water and followed by distilled water. The neck and stopper were wrapped by a brown paper with the help of rubber bands, sterilized the sample bottles in oven for 2 hours.

(ii) Petri–plates:

Plates washed with washing soda as well as with chromic acid and rinsed with clean water and then sterilized in oven at 160°C to 180°C for 2–3 hours.

(iii) Test tubes:

The test tube were washed and rinsed with water. Now plugged the tube with non–acidic and than autoclaved.

(iv) Pipettes:

Pipettes were washed and fitted with plug of cotton at the upper end and wrapped in paper then sterilized in oven for 2–3 hours.

(b) Sample Collection:

Water sample was taken from 6″ below the surface of waste water by keeping and opening the mouth of container against the flow of water.

(i) Sampling for DO:

For DO determination the samples were collected in a clean 300 ml BOD bottles. The bottles were filled completely and the sample preserved by the addition of 2 ml each of $MnSO_4$ and alk. KI at the sampling spot.

(ii) Sampling for pH, Total Hardness, Total Alkalinity, BOD, COD, Chlorides, Total Solids:

The samples were collected in plastic jerriken of 2 litre capacity. These were cleaned as in the case of glass wares. The cleaned jerriken was washed 2–3 times with water for sample collection.

(iii) Sampling for Microbiological Analysis:

Clean and sterilized sampling bottle was used. It was held at its base by hand and dip into the water with its neck downward upto 6″ below the water surface and then its mouth was opened by removing its stopper against the current. The bottle was not filled completely. The mouth of the bottle was closed under water by replacing stopper to avoid contamination of the sample. The neck and stopper were wrapped by a brown paper with the help of rubber bands.

(C) Water Analysis Methods

The physico–chemical and microbiological analysis of water samples made by following standard methods outlined by A.P.H.A. (1980) and Khanna (1993).

Samples were analysed for following parameters.

(a) Physico–Chemical Parameters :

(1) Temperature

(2) pH

(3) Total Solids (TS)

(4) Total Hardness (TH)

(5) Total Alkalinity (TA)

(6) Chlorides (Cl^-)

(7) Free Carbon Dioxide

(8) Dissolved Oxygen (DO)

(9) Biochemical Oxygen Demand (BOD)

(10) Chemical Oxygen Demand (COD)

(b) Microbiological Parameters:

(1) Standard Plate Count (SPC)

(2) Most Probable Number (MPN) for total coliforms.

Methodology

Physico–Chemical Parameters

1. Temperature

Temperature of waste water was measured using centigrade thermometer. The bulb of thermometer was immersed in water about 6" below the surface and the temperature noted.

2. Hydrogen Ion Concentration (pH)

pH is the $-\log^{10}$ of hydrogen ion concentration in the solution. The determination of pH is very important control test for evolution of the solution for under examination.

Procedure:

pH of the waste water samples was determined by paper strip method. For this purpose a paper strip which has already been soaked in sensitive compound was dipped in effluent sample for nearly 3–5 seconds and taken out allowed to dry. The colour so developed was matched with standard colour range provided with it and pH was noted directly.

3. Total Solids (TS)

Total Solids were determined as the residue left after evaporation of unfiltered sample. It indicates the quantity and presence of solid pollutants dissolve or undissolved in water.

Procedure:

A porcelain dish was used for evaporation. First of all, the dish was heated and then cooled and weighed. Now 100 ml of unfiltered sample taken in dish. It was placed in hot air oven operating at a temperature of 105°C and unfiltered sample in the dish was evaporated at same temperature. Final weight of the dish was taken after complete evaporation.

Calculation:

Total solids were calculated using the following formula.

$$\text{Total Solids (mg/l)} = \frac{(X-Y) \times 1000}{V}$$

where,

X – final weight of the dish
Y – initial weight of the dish
V – Volume of water sample taken (100ml)

4. Total Hardness (TH)

Hardness is the property of water which prevents the lather or foaming formation with soap and increase the boiling point of water. Principal cations imparting Hardness are calcium and magnesium. The anions are responsible for hardness are mainly bicarbonates, carbonates, sulphate, chloride, nitrate silicates etc. Hardness is called temporary if it is caused by the presence of bicarbonate and carbonate salts of cations. Permanent hardness is caused mainly by sulphate and chlorides of metals.

Reagents:

(a) EDTA Solution (0.01):

Dissolve 3.723 gm of disodium salt of EDTA in distilled water to prepare one litre of solution store in Pyrex bottle.

(b) Buffer solution:

Dissolve 6.75 gm of Ammonium chloride in 75 ml of liquid Ammonia and dilute to 100 ml with distilled water.

(c) Erichrome–Black–T–indicator:

Dissolve and mix 0.5 gm of Erichrome Black–T with 100 gm NaCl (A.R.).

Procedure:

Take 50 ml of sample in a conical flask. Add 2.3 ml of buffer solution now add 0.2–0.3 ml of indicator the solution turns wine red. Titrate the contents with EDTA solution colour changes from wine red to blue.

Calculation:

$$\text{Hardness as mg/l as } CaCO_3 = \frac{\text{ml of EDTA used} \times 1000}{\text{ml of sample used}}$$

5. Total Alkalinity

Alkalinity of a water is a measure of its capacity to neutralise acids, without significant change in pH. Alkalinity is measured by titrating a sample with 1N H_2SO_4 solution. For highly alkaline samples the first step to titration at a pH of 8.3. The second phase of, first in case of water with an initial pH of less than 8.3 is titrating to an indicated pH of 4.3. Calorimetric indicators, chemicals that change the colour of specific pH values, can be used to determine the end point of these titrations if a pH meter is not used. Phenolphthalein turns from pink to colourless at pH 8.3 and methyl orange changes from orange to pink. The part of the alkalinity above pH 8.3 is called as phenolphthalein alkalinity.

Reagents:

(a) Standard sulphuric acid (1N) :

49 gm of sulphuric acid in one litre of distilled water.

(b) Phenolphthalein indicator:

Dissolve 0.5 g of phenolphthalein in 50 ml of 95% ethanol and add 50 ml of distilled water. Add 0.05 N CO_2 free NaOH solution drop wise until the solution become faintly pink.

(c) Methyl orange indicator:

Dissolve 0.5 g of methyl orange in 100 ml of distilled water.

Procedure:

(1) We take 50 ml of sample in erlenmayer flask and add 2–3 drops of phenolphthalein indicator. If the solution remain colourless then phenolphathalein alkalinity is absent. If it turns pink then titrate it with standard sulphuric acid till the solution become colourless.

(2) Take 50 ml of samples add 2–3 drops of methyl orange indicator. If the solution turns pink alkalinity is absent. If it turns orange, titrate it with 1N standard sulphuric acid till colour change from orange to pink.

Calculation:

$$\text{Alkalinity (mg/l)} = \frac{(\text{ml} \times \text{N}) \text{ of standard acid} \times 50}{\text{ml of sample}}$$

6. Chlorides

Silver nitrate reacts with chlorides to form very lightly soluble ppt of AgCl. Chloride test may be performed to measure the efficiency of chlorination of effluents. At the end point when all the chloride get precipitated, free silver ions react with chromate to form silver chromate of reddish brown colour.

Regents:

(a) Silver nitrate ($AgNo_3$):

(0.025N) 3.40 gm of fried silver nitrate is dissolved in distilled water to make one litre of solution and keep in dark bottles.

(b) Potassium Chromate (K_2CrO_4):

5.0 gm of potassium chromate was dissolved in 100 ml of distilled water.

Procedure:

Take 50 ml of sample in a erlenmeyer flask and add 2 ml of potassium chromate (5%) solution. Titrate the contents against 0.025 N silver nitrate until a persistent reddish brown thing appears.

Calculation:

$$\text{chloride (mg/l)} = \frac{\text{ml of } AgNO_3 \times 1000 \times 35.5}{\text{ml of sample}}$$

7. Free Carbondioxide

Free CO_2 reacts with sodium carbonate sod. hydroxide to form sodium bicarbonate. The completion of reaction is indicated by the development of pink colour. Characteristics of phenolphthalein indicator at the equivalence pH of 8.3. A0.01 N sodium bicarbonate solution containing recommended value of phenolphthalein indicator is a suitable colour until similarity is obtained with the colour at the end point.

Reagents:

(a) Sodium Hydroxide (0.05N):

Prepare 1N Sod. hydroxide by dissolving 40 gm of NaOH in Co_2 free distilled water to make 1 litre of solution. Dilute 50 ml of 1.0N NaOH to one litre. Standardised it with sulpuhric acid.

(b) Phenolphthalein indicator:

Dissolve 0.5 gm of phenolphthalein in 50 ml of ethanol (95%) and 50 ml of distilled water. Add 0.05N carbon dioxide free NaOH solution, drop wise until the solution just turns faintly pink.

Procedure:

Take 50 ml of sample in a erlenmeyer flask and few drops of phenolphthalein indicator is added to it. If the colour turns pink free CO_2 is absent. If the sample remain colourless, titrate it against 0.05N NaOH to the end point till a pink colour appears.

Calculation:

$$\text{Free } Co_2 \text{ (mg/l)} = \frac{(\text{ml} \times \text{N}) \text{ of NaOH} \times 1000 \times 44}{\text{ml of sample}}$$

8. Dissolved Oxygen (DO)

Dissolved oxygen is the measure of oxygen concentration in the given water sample. Visually DO is determined by two methods.

(i) Winkler's Iodometric method.

(ii) Azide modification of Winkler's method.

Although there is a minute difference between two methods but the second method was followed for DO determination.

Azide modification of Winkler's method, the magnous sulphate reacts with alkaline KOH or NaOH to form white ppt. of magnous hydroxide, which in presence of oxygen get oxidized into brown colour oxide (MnO_2). In the strong acidic medium, maganese ions are reduced by Iodide ions which gets converted to iodine equivalent to the original concentration of oxygen in sample. The iodine can be titrated against sulphate using starch as an indicator. Certain oxidizing and reducing materials may effectively interfere with the determination of oxygen by converting iodine ions. Azide modification removes such substances especially nitrate is quite suitable for polluted water.

Reagents:

(a) Sodium Thiosulphate Solution (0.025N):

Dissolve 24.82 gm of sodium thiosulphate in boiled distilled water and make upto to one litre solution. Add 0.4 gm Borax as stabilizer dilute the 0.1N stock solution 4 time means (250ml into 1000ml).

(b) Alkaline potassium Iodine solution:

Dissolve 100 gm of KI in 200 ml of distilled water.

(c) Magnous sulphate solution:

100 gm of magnous sulphate is dissolved in 200 ml of distilled water and filtered.

(d) Starch Indicator:

5 gm of starch is dissolved in 100 ml of distilled hot water and few drops of formaldehyde solution were added.

(e) Sulphuric Acid:

Concentrated sulphuric acid of specific gravity (1.84) is used.

Procedure

Take the known volume of sample 300 ml in a glass stopper bottle (BOD bottle) and add 1–2 ml of each $MnSO_4$ and alkaline KI solution. A ppt will develop. Shake the solvent by inverting the bottle and kept for some time to settle down the ppt. Add 1–2 ml of conc. $H2SO_4$ and shake well the contents to dissolve the ppt.

Now a known amount of sample 50 ml was poured from BOD bottle to conical flask. Add few drops of starch indicator and titrate the content with 0.025N sodium thiosulphate solution, till the blue colour appeared by the addition of starch indicator disappeared.

Calculation:

If 0.025N sodium thiosulphate was used to measure the dissolve oxygen in a volume equal to 50 ml of original sample. 1 ml of titrant is equal to the 1.0 mg/l dissolve oxygen.

$$\text{DO 9mg/l)} = \frac{(\text{ml} \times \text{N}) \text{ of titrant} \times 8 \times 1000}{V_2 \dfrac{V_1 - V}{V_1}}$$

where

V_1 = Total initial volume of sample in bottle.
V_2 = Volume of titrated sample.
V = Volume of $MnSO_4$ and KI added.

9. **Bio–chemical oxygen demand (BOD)**

BOD is the measure of organic matter present in a water sample and can be defined as the amount of oxygen required by the micro–organism, in stabilizing the biologically degradable organic matter under aerobic condition. In other words, BOD is the quantity of oxygen utilised by a mined population of micro–organism in the aerobic oxidation of organic matter in a sample of waste water at a temperature of 20°C.

The principle of the method involves, measuring difference of oxygen concentration of the sample before and after incubation of 5 days at 20°C.

Reagents:

(a) Sodium thiosulphate (0.025N)

(b) Alkaline potassium idoide solution

(c) Magnous sulphate solution

(d) Starch indicator

(e) Sulphuric Acid (1.84 specific gravity)

Method of preparation of above reagents is same as in DO Test.

(f) Phosphate Buffer:

Dissolve each 85 gm of potassium dihydrogen phosphate, 21.75 gm of Dipotassium hydrogen phosphate, 33.4 gm of Disodium hydrogen phosphate and 21.7 gm of NH_4Cl in distilled water to make the volume one litre and pH should be adjusted to 7.2.

(g) Magnesium sulphate:

Dissolve 82.5 gm of $MnSO_4.H_2O$ to one litre of distilled water.

(h) Ferric chloride solution:

Dissolve 0.25 gm of $FeCl_3.6H_2O$ to one litre of distilled water.

Procedure:

First of all dilution of water was prepared in a glass container by bubbling compressed air in good quality distilled water for 24 hours. Add 1 ml each of phosphate buffer, magnesium sulphate, calcium chloride and ferric chloride solution for each litre of dilution water and mined thoroughly. Neutralize the sample to pH 7.0 by using 1N NaOH since DO in the sample is likely to be exhausted. Now record the DO in diluted sample and then keep the same dilution of sample in BOD bottles for 5 days in incubator operating at temperature 20°C. After 5 days sample is again titrated for DO content.

Also use a dilution water blank as a rough check on the quality of unseeded dilution water and cleanliness of incubation bottles together with each batch of samples. Determined initial and final DO.

Calculation:

$$\text{BOD (mg/l)} = \frac{(D_0 - D_5) - (B_0 - B_5)}{P}$$

Where

D_0 = Initial DO of diluted waste water sample about 15 minutes after preparation.

D_5 = Final DO of diluted waste water sample after incubation for 5 days.

B_0 = Initial DO of blank.

B_5 = Final DO of blank after incubation of 5 days.

P = Decimal fraction of the waste water sample used.

$$= \frac{\text{ml. of waste water sample used}}{\text{ml of volume of BOD bottle}}$$

10. Chemical Oxygen Demand (COD)

COD is a measure of oxygen consumed during the oxidation of oxidizing agent. The sample is refluxed with $K_2Cr_2O_7$ in presence of mercuric sulphate to neutralize the effects of chloride and silver sulphate. The excess of $K_2Cr_2O_7$ is titrated against. Ferrous Ammonium sulphate (F.A.S.) using Ferrion as an indicator. The amount of $K_2Cr_2O_7$ is used in proportional to the oxidizable organic matter present in the sample.

Reagents:

(a) Potassium Dichromate solution ($K_2Cr_2O_7$ 0.025N):

Dissolve 12.25 gm of dried $K_2Cr_2O_7$ in one litre distilled water to make 0.025N solution. Dilute it to 10 times (100 to 1000 ml).

(b) Ferrous Ammonium Sulphate (F.A.S.) (0.1N):

Dissolve 39.2 gm of $Fe(NH_4)_2(So_4)_2.6H_2O$ in distilled water adding 20 ml of cone. H_2SO_4 to make one litre of solution.

(c) Ferrous Ammonium Sulphate (0.01N):

Dilute the 0.1N Ferrous Annomium Sulphate solution to 10 times (100 to 1000 ml).

(d) Ferrion indicator : as such

(e) Sulphuric Acid (specific gravity 1.84) : as such

(f) Mercuric Sulphate : $HgSO_4$ solid as such

(g) Silver sulphate : Ag_2SO_4 solid as such

Procedure:

Take 20 ml of sample in a COD flask and add 10–15 ml of 0.025 potassium dichromate solution. Extreme care should be taken in case of low COD. Add a pinch of Ag_2SO_4 and $HgSO_4$ to remove the effects of chlorides. Now add 30 ml of $HgSO_4$. Reflux the contents for 2 hours on a water bath cool the flask and add distilled water to made the final volume to about 140 ml. Add 2–3 drops of ferrion indicator mix thoroughly and titrate with 0.01 N Ferrous Ammonium Sulphate. Run a blank too with and distilled water using same quantities of chemicals.

Calculation:

$$\text{COD (mg/l)} = \frac{(b-a) \times \text{N of } K_2Cr_2O_7 \times 1000 \times 8}{\text{ml. of sample}}$$

where

a = ml of titrant with sample

b = ml of titrant with blank.

Microbiological parameters

1. **Standard Plate Count (SPC)**

In solid medium, counting of organism depends upon the fact that living cell will proceed to multiply and in time will produce sufficient progeny to form a colony visible with nacked eyes. Since bacteria in water as single cells, pairs, groups or chains and every individual living cell could develop in to a separate colony on incubation.

Therefore no colonies appearing on a plate does not necessarily represent the total number of organisms present on test volume but gives an overall estimate of bacteria present in water. The results are expressed as the number of colonies per ml.

(a) Preparation of serial dilution:

5–6 culture tubes were filled with 9 ml of distilled sterilized water in each. The tubes were cotton plugged and sterilized at 15 lbs. for 10–15 minutes in an autoclave. Shake the sample vigorously and transfer one of sample to the test tube and after proper shaking 1 ml from first tube to second tube and so on. Use separate sterilized pipettes for each transfer and after all 1 :10, 1 :100, 1 :1000; 1 :10000, 1 :100000 dilution were made.

(b) Preparation of Medium:

Standard plate count utilizes 'Nutrient Agar Medium' (NAM). The composition of medium is as follows.

Peptone	–	5.0 gm
Beef Extract	–	3.0 gm
Agar – agar	–	18.0 gm
Distilled water	–	1000 ml
pH	–	7.5

For preparation of the medium 5 gm of peptone and 3 gm of Beef Extract were dissolved in 500 ml of hot distilled water. In another 500 ml of distilled water 18 gm of agar–agar was dissolved by heating. Both the solutions were mixed homogeneously, heated and poured in conical flask of appropriate sizes made of Borosil. Cotton plugged and were sterilized at 15 lbs for 15 minutes in an autoclave.

Procedure:

Sterilize 5 pairs of Borosil petriplates in the oven. Transfer 1 ml of serially diluted sample from the last dilution to the petriplate by opening the plate gently from one side. 15 ml of NAM is poured to these plates. The temperature of medium at pouring time should not exceed 40 to 42°C. Mix the medium with inoculum by gently shaking. Wait till the medium solidity. Invert the plates as the medium solidifies. Incubate these plates for 24 hours at 37°C temperature. Count the colonies appeared after the incubation period with the help of colony counter and apply the formula to calculate the SPC/ml.

Calculation:

$$\text{SPC/ml} = \frac{\text{colonies counted}}{\text{dilution factor}}$$

2. Most Probable Number (MPN)

The coliform group comprises all aerobic facultative anaerobic, Gram negative, non–spore forming rod shaped bacteria that ferment lactose with gas production within 48 hours at 38°C. The coliform group test clearly indicate the fecal pollution of water. This is also most useful test for drinking water. This test includes presumptive, conformed and completed test.

Presumptive Test:

'McConkey's' Broth is the required media for this test. The composition of the media is as follows

Bile salts (sodium taurocholate)	–	5.0 gm
Peptone	–	20.0 gm
Lactose	–	10.0 gm
NaCl	–	5.0 gm
Bromocresol purple	–	1.0 ml
Distilled water	–	1000 ml
pH	–	7.4

Procedure:

Prepare serial dilution of the sample same as in standard plate count, take 5 fermentation tubes for each dilution. Pour these tubes with 9 ml of broth and but one Derham's tube to each fermentation tube in inverted position. Now plug the mouth of all the tubes with cotton plugs. Now sterilize all the tubes in Autoclave.

Shake all the dilution vigorously before moving the sample to the fermentation tubes. Add one ml of the diluted sample to germentation tubes. Separate sterilised pipettes were used to different dilutions shake fermentation tube gently. Now incubate tubes for 48 hours at 35–37°C. Examine all the tubes after 24–48 hours for gas production. Record the tubes as positive (+ve) showing gas production in Derham's tube even a tinny bubble should also be recorded as positive test. This test is now subject to the confirmatory test. Incubate negative showing result tubes for further 24 hours.

Confirmatory Test:

For this test Brilliant Green Lactose Broth (BGLB) is given below:

Peptone	–	10.0 gm
Lactose	–	10.0 gm
Oxagall	–	20.0 gm
Brilliant green	–	13.3 mgm
Distilled water	–	1000 ml
pH	–	7.2

Procedure:

Prepare fermentation tubes with 10 ml. BGLB medium with Derham's tube the number of tubes should be equal to all positive tubes equal to in presumptive test. Shake positive tubes gently. Now transfer one loopful of liquid to the tubes having BGLB broth with the help of inoculation needle. Incubate these tubes for 48 hours at 35–37°C. Reward the tubes as positive which showing gas production. Put the number of positive tubes in the MPN table and then in the formula to get the MPN/100 ml. To confirm this test completed test should done.

Completed Test:

For this test Eosin Methylene Blue Agar (EMB) medium is required. The composition of EMB is as given below:

Peptone	–	10.0 gm
Lactose	–	10.0 gm
K_2HpO_4	–	2.0 gm
Eosin 'Y'	–	0.4 gm
Methylene blue	–	0.064 gm
Agar–agar	–	15.0 gm
Distilled water	–	1000 ml
pH	–	7.1

Procedure:

Prepare EMB agar medium petriplates. The number of petriplates should be equal to the number of positive tubes in confirmatory test. Streak the inoculum from positive B.G.L.B. tubes. After this plates are incubated for 24 hours at 35–37°C. Examine the plate for bacterial colony appearance colonies with dark centre will be the typical coliform colony. Pink opaque and inoculated colonies (a typical colonies) may also belong to the coliform groups other colonies should discarded.

Now isolate the colony of coliform and inoculate in the 'McConkey's Broth and Examine for gas production. (A repetition of the presumptive test but with the specific coliform colony). Also examine the colonies by Gram's staining. For this transfer the colonies to a NAM slant subject the colonies from Agar slant to Gram's staining. It organism appear rod (bacilli) shaped, red stained, occurring singly or in pairs or in short chains. The test is completed to indicate the presense of coliform bacteria in sample provided.

Observation

The result obtained from the observation made during December 1995 to April 1996, on the various physico–chemical and microbiological parameters included in this study shown in Table 1.1 to 1.3.

1. **Temperature**

   The minimum value of temperature among all sampling sites was recorded 16.3°C during December while maximum value was found 24.2°C in sugar mill in April.

(i) IDPL effluents

The minimum temperature was recorded 17.5°C during December and maximum temperature was observed 22.0°C in April.

(ii) BHEL effluents

The minimum temperature was observed 18.0°C in December and maximum temperature recorded was 23.0°C during April.

(iii) Sugar mill effluents

The minimum value of temperature was noted 16.3°C in December and the maximum value was recorded 24.2°C in April.

2. Total Solids

The minimum value of total solids among all sampling stations was recorded 3156.0 mg/l in IDPL waste during February while maximum value was found 9357.0 mg/l in sugar mill in April.

(i) IDPL effluents

Total solids was observed minimum 3156.0 mg/l during February and it was noted maximum 4250.0 mg/l during April.

(ii) BHEL effluents

The minimum total solids was recorded 5904.0 mg/l in April. While the maximum value was noted 7002.0 mg/l in February.

(iii) Sugar mill effluents

The minimum total solids was observed 8155.0 mg/l and maximum was found 9357.0 mg/l during December and April.

3. pH

The minimum value of pH among all industries was recorded 6.5 in BHEL during December. While maximum value was found 7.8 in IDPL in April.

(i) IDPL effluents

The minimum value of pH was noted 7.0 in December and maximum was 7.8 during April.

(ii) BEHL effluents

The minimum and maximum pH was recorded 6.5 and 7.0 during December and April respectively.

(iii) Sugar mill effluents

The minimum pH was observed 7.1 during December and maximum value was noted 7.7 during April.

4. Total Hardness (TH)

The minimum value of total hardness among all sampling sites was recorded 210.0 mg/l in IDPL during December. While maximum value was found 370.6 mg/l in sugar mill in April.

(i) IDPL effluents

The minimum value of total hardness was noted 210.0 mg/l in December and maximum was recorded 235.7 mg/l during April.

(ii) BHEL effluents

The minimum value of total hardness was noted 265.2 mg/l in December and maximum was recorded 295.0 mg/l during April.

(iii) Sugar mill effluents

The minimum total hardness was found 360.0 mg/l and maximum was observed 370.6 mg/l during December and April respectively.

5. Total Alkalinity

The minimum value of total alkalinity among all industries was recorded 115.8 mg/l in IDPL during December. While maximum value was found 214.7 mg/l in sugar mill in April.

(i) IDPL effluents

The minimum total alkalinity was observed 115.8 mg/l

during December and maximum was noted 132.5 mg/l in April.

(ii) BHEL effluents

The total alkalinity was found minimum 140.0 mg/l while the maximum observed 165.3 mg/l in December and April.

(iii) Sugar mill effluents

The total alkalinity was recorded minimum 168.0 mg/l in December and it was noted maximum 214.7 mg/l during April.

6. **Chlorides (Cl)**

The minimum value of chlorides among all sampling stations was recorded 13.5 mg/l in sugar mill during December while maximum value was found 120.7 mg/l in IDPL in April.

(i) IDPL effluents

The minimum chloride was found 112.0 mg/l during February and maximum value was noted 120.7 mg/l during April.

(ii) BHEL effluents

The minimum value of chlorides was recorded 23.0 mg/l and maximum was 34.6 mg/l during December and April.

(iii) Sugar mill effluents

The minimum value was recorded 13.5 mg/l during December and whereas the maximum value was found 18.1 mg/l in April.

7. **Dissolved Oxygen (DO)**

The minimum value of dissolved oxygen among all the sampling sites was recorded 0.8 mg/l in sugar mill in April while maximum value was found 6.5 mg/l in IDPL in December.

(i) IDPL effluents

The minimum and maximum value was noted 4.1 mg/l and 6.5 mg/l during April and December respectively.

(ii) BHEL effluents

Minimum and maximum value of dissolved oxygen was

recorded 0.9 mg/l and 2.1 mg/l during April and in December.

(iii) Sugar mill effluents

The minimum value of dissolved oxygen was recorded 0.8 mg/l in April and where as the maximum value was found 1.8 mg/l during December.

8. **Bio–Chemical Oxygen Demand (BOD)**

The minimum value of bio–chemical oxygen demand among all industries was recorded 95.0 mg/l in IDPL during December while maximum was found 405.0 mg/l in Sugar mill in April.

(i) IDPL 3effluents

The minimum value of bio–chemical oxygen demand was noted 95.0 mg/l in December and maximum was 125.5 mg/l during April.

(ii) BHEL effluents

The minimum and maximum bio–chemical oxygen demand was recorded 325.0 mg/l and 398.6 mg/l during December and April respectively.

(iii) Sugar mill effluents

The minimum bio–chemical oxygen demand was observed 358.1 mg/l during December and maximum value was noted 405.5 mg/l during April.

9. **Chemical Oxygen Demand (COD)**

The minimum value of chemical oxygen demand among all sampling sites was recorded 120.3 mg/l in IDPL during December while maximum value was found 609.0 mg/l in sugar mill in April.

(i) IDPL effluents

The minimum chemical oxygen demand was noted 120.3 mg/l during December and maximum value was noted 138.1 mg/l during April.

(ii) BHEL effluents

The minimum value of chemical oxygen demand was recorded 440.5 mg/l during December and while the maximum was found 560.9 mg/l in April.

(iii) Sugar mill effluents

The minimum value of chemical oxygen demand was noted 579.2 mg/l and where as the maximum value was recorded 609.0 mg/l during December and April.

10. Free CO_2

The minimum value of free CO_2 among all sites was recorded 7.5 mg/l in IDPL during December while maximum value was found 26.5 mg/l in sugar mill in April.

(i) IDPL effluents

The minimum free CO_2 noted 7.5 mg/l during December and while the maximum value was found 10.7 mg/l in April.

(ii) BHEL effluents

Minimum and maximum value was observed 13.9 mg/l and 17.0 mg/l during December and April.

(iii) Sugar mill effluents

The minimum value of free CO_2 was found 18.2 mg/l in December and whereas the maximum value was noted 26.5 mg/l during April.

11. Standard Plate Count (SPC)

The minimum value of standard plate count among all industries was recorded 44×10^5/ml in IDPL during february while maximum value was found 67×10^5 in sugar mill in February.

(i) IDPL effluents

The minimum value of SPC was recorded 44×10^5/ml during February while the maximum 55×10^5/ml during April.

(ii) BHEL effluents

The minimum value was observed 54×10^5/ml during December and whereas the maximum 62×10^5/ml in April.

(iii) Sugar mill effluents

The minimum value of SPC was noted 52×10^5/ml during April while the maximum value was recorded 67×10^5/ml in February.

12. Most Probable Number (MPN)

The minimum value of most probable number among all sampling sites was recorded 21/100 ml in BHEL during December while maximum value was found 44/100 ml in sugar mill in April.

(i) IDPL effluents

The minimum MPN recorded 23/100 ml during December, while the maximum was 42/100 ml in April.

(ii) BHEL effluents

The minimum and maximum was noted 21/100 ml and 35/100 ml during December and April respectively.

(iii) Sugar mill effluents

The minimum value of MPN was observed 29/100 ml in December and whereas the maximum value was noted 44/100 ml during April.

Table 1.1: Physico–chemical and Microbiological parameters of IDPL effluents

Parameters		*Months*			*Average*	
		Dec.	*Feb.*	*April*		
Physico–chemical parameters (mg/l)	temp (°C)	17.5	19.0	22.0	19.5	± 2.29
	TS (mg/l)	3902.0	3156.0	4250.0	3769.3	± 558.93
	pH	7.0	7.3	7.8	7.3	± 0.212
	TH	210.0	222.2	235.7	222.6	± 12.86
	TA	115.8	222.2	132.5	122.1	± 9.07
	Cl^-	114.2	112.0	120.7	115.6	± 4.52
	DO	6.5	5.2	4.1	5.2	± 1.20
	BOD	95.0	106.5	125.5	109.0	± 15.40
	COD	120.3	132.6	138.1	130.3	± 9.11
	Free CO_2	7.5	10.0	10.7	9.4	± 1.68
Microbiological parameters	SPC (10^5/ml)	48.0	44.0	55.0	49.0	± 5.57
	MPN (/100 ml)	23.0	34.0	42.0	33.	± 9.54

Table 1.2: Physico–chemical and Microbiological parameters of BHEL effluents

Parameters		*Months*			*Average*
		Dec.	*Feb.*	*April*	
Physico–chemical parameters (mg/l)	temp (°C)	18.0	19.5	23.0	20.3 ± 2.79
	TS (mg/l)	670.0	7002.0	5904.0	6535.3 ± 567.22
	pH	6.5	6.8	7.0	6.8 ± 0.252
	TH	265.2	280.9	295.0	280.4 ± 14.91
	TA	140.0	142.5	165.3	149.2 ± 13.94
	Cl^-	23.0	33.8	34.6	30.4 ± 6.48
	DO	2.1	1.2	0.9	1.4 ± 0.625
	BOD	325.0	393.0	398.6	372.2 ± 40.97
	COD	440.5	543.2	560.9	514.8 ± 65.01
	Free CO_2	13.9	16.2	17.0	15.7 ± 1.61
Microbiological parameters	SPC (10^5/ml)	54.0	59.0	62.0	58.3 ± 4.04
	MPN (/100 ml)	21.0	28.0	35.0	28.0 ± 7.00

Table 1.3: Physico–chemical and Microbiological parameters of Sugar mill effluents

Parameters		*Months*			*Average*
		Dec.	*Feb.*	*April*	
Physico–chemical parameters (mg/l)	temp (°C)	16.3	18.0	24.2	19.6 ± 4.16
	TS (mg/l)	8155.0	8262.0	9357.0	8591.3 ± 662.24
	pH	7.1	7.5	7.7	7.4 ± 0.306
	TH	360.0	368.3	370.6	366.3 ± 5.58
	TA	168.0	191.3	214.7	191.3 ± 23.35
	Cl^-	13.5	17.2	18.1	16.2 ± 2.44
	DO	1.8	1.0	0.8	1.2 ± 0.530
	BOD	358.1	382.0	405.5	381.8 ± 23.70
	COD	579.2	590.1	609.0	592.7 ± 15.08
	Free CO_2	18.2	23.1	26.5	22.6 ± 4.17
Microbiological parameters	SPC (10^5/ml)	60.0	67.0	52.0	59.6 ± 7.51
	MPN (/100 ml)	29.0	35.0	44.0	36.0 ± 7.5500

Discussion

Pollution is the greatest crime of mankind against himself, with the rapid pace of industrialisation. Effluents have posed a serious threat to the vast and varied resources of the country. Water quality of major river system is getting rapidly degraded due to massive discharges of municipal waste and industrial waste of diverse origin.

In the light of above effective management and control of water pollution and becoming increasingly important for sustainable development and human welfare, because industrial effluent discharged into water bodies are responsible for a number of mortalities and incapacitations in the world.

The characters of effluent studies included temperature, pH, Total Solids, Total Hardness, Total Alkalinity, Chlorides, Dissolved Oxygen, Bio–chemical Oxygen Demand, Chemical Oxygen Demand, Free Co_2, Standard Plate Count, Most Probable Number. Sampling was done from December 1995 to April 1996 by months.

The effluents of all industries showed low values of DO and high values of BOD, COD, SPC, MPN and Total Solids. The DO and BOD values showed an inverse relationship.

In the present work, industrial effluents from three different industries viz. IDPL (Indian Drugs and Pharmaceuticals Ltd.) BHEL (Bharat Heavy Electricals Ltd.) and RBNS sugar mill were selected for the study of their physico–chemical and micro–biologicial parameters.

The comparative annual average of the effluent of three industries for their different parameters are given in Table 1.4.

The result of study bring to light that temperature of the industrial effluents increased from December to April. The maximum average temperature was recorded 20.3^{o}C ± 2.79 in the effluents of BHEL and the minimum average temperature recorded was 19.5^{o}C ± 2.29 in the effluents of IDPL.

It evident that the dissolved oxygen (DO) values of industrial effluents decreased from December to April. The maximum dissolved oxygen was recorded 5.2 mg/l ± 1.20 in the effluents of IDPL. The minimum average DO recorded was 1.2 mg/l ± 0.530 in effluents of sugar mill. The results also supported by the work of *Horne et al.* (1987).

Table 1.4: Comparative Study of Physico–chemical and Microbiological parameters of three different industrial effluents.

Parameters		Sites		
		IDPL	BHEL	Sugar Mill
	temp (°C)	19.5 ± 2.29	20.3 ± 2.79	19.6 ± 4.16
	TS	3769.3 ± 558.93	6535.3 ± 567.22	8591.3 ± 662.24
	pH	7.3 ± 0.212	6.8 ± 0.252	7.4 ± 0.306
	TH	222.6 ± 12.86	280.4 ± 14.91	366.3 ± 5.58
	TA	122.1 ± 9.07	149.2 ± 13.94	191.3 ± 23.35
	Cl	115.6 ± 4.52	30.4 ± 6.48	16.2 ± 2.44
	DO	5.2 ± 1.20	1.4 ± 0.625	1.2 ± 0.530
	BOD	109.0 ± 15.40	372.2 ± 40.97	381.8 ± 23.70
	COD	130.3 ± 9.11	514.8 ± 65.01	592.7 ± 15.08
	Free CO_2	9.4 ± 1.68	15.7 ± 1.61	22.6 ± 4.17
Micro–biological	SPC (10^5/ml)	49.0 ± 5.57	58.3 ± 4.04	59.6 ± 7.51
	MPN (/100 ml)	33.0 ± 9.54	28.0 ± 7.0	36.0 ± 7.55

The result of study bring to light that temperature of the industrial effluents increased from December to April. The maximum average temperature was recorded 20.3°C ± 2.79 in the effluents of BHEL and the minimum average temperature recorded was 19.5°C ± 2.29 in the effluents of IDPL.

It evident that the dissolved oxygen (DO) values of industrial effluents decreased from December to April. The maximum dissolved oxygen was recorded 5.2 mg/l ± 1.20 in the effluents of IDPL. The minimum average DO recorded was 1.2 mg/l ± 0.530 in effluents of sugar mill. The results also supported by the work of *Horne et al.* (1987).

It is evaluated that the bio–chemical oxygen demand (BOD) values of industrial effluents increased from December to April. The maximum average BOD was noted 381.8 mg/l ± 23.70 in the effluents of sugar mill and minimum avareage value was recorded 109.0 mg/l ± 15.40 in effluents of IDPL.

It may be noted that during this period i.e. from December 1995 to April 1996 there was corresponding increase in termperature also indicating that the increase in temperature showed increase in BOD and corresponding decrease in DO values. So it is stated that the temperature and DO shows inverse relationship. While direct relationship occur between temperature and BOD values.

Bhargava (1985) also recorded higher BOD values during summer when the temperature was relatively higher. He attributed it to increased bacterial activity at relatively higher temperature. These findings were also supported by the work of Kamath (1985), Abeliovich (1985), and Riedel *et al.* (1990) who found that, where heated effluents were discharged into receiving water bodies the DO values of such water bodies found very low and BOD values relatively higher.

It is evident that the value of free CO_2 increased from December to April. The maximum average free CO_2 was recorded 22.6 mg/l ± 4.17 in effluents of sugar mill and the minimum average free CO_2 was recorded 9.4 mg/l ± 1.68 in IDPL effluents.

These findings were also supported by Balasu (1966) and Singh *et al.* (1975).

It is evident that the values of chemical oxygen demand (COD) were increased from December to April. The maximum average value of COD noted was 592.7 mg/l ± 15.08 in the effluents of sugar mill and minimum average value recorded was 130.3 ± 9.11 in IDPL effluents.

According to Camp and Meserv (1974) COD values do not differentiate between stable and unstable organic pollution therefore it is more than BOD values. The results were also supported by Gerrote *et al.* (1995). In all the industrial effluents it was noted that COD values are higher than BOD values. Above notification also supported by the work of Kim *et al.* (1990).

It is noted that the value of Hydrogen ion concentration (pH) increased from December to April in all the sampling sations. The maximum average value recorded was 7.4 ± 0.306 of sugar mill effluents and minimum average value was 6.8 ± 0.252 in BHEL effluents.

It is evident that the total hardness (TH) increases from December to April and maximum average hardness was 366.3 mg/l ± 5.58 in effluents of sugar mill and minimum average value recorded was 222.6 mg/l ± 12.86 in IDPL wastes.

Indian water works association clssified the hardness of water in following types:

6–75	Soft
75–150	Moderately hard
150–300	Hard
Above 300	Very hard

So it was concluded that the industrial effluents come under the hard and very hard categories.

The result of the study shows that the total solids (TS) increases from December to April. The maximum avareage value was 8591.3 mg/l ± 665.24 in sugar mill effluents. While minimum average value recorded was 3769.3 mg/l ± 558.93 in IDPL wastes. Results also supported by the Zahid *et al.* (1990) in biological filter effluents studies.

It is evident that the value of total alkalinity (TA) increases from December to April. The maximum average value noted was 191.3 mg/l ± 23.35 in sugar mill wastes and minimum average value recorded was 122.1 mg/l ± 9.07 in IPDL effluents.

The study reveals that chloride (Cl) concentration increases from December to April. The maximum average value recorded was 115.6 mg/l ± 4.52 in IDPL wastes and minimum average value was 16.2 mg/l ± 2.44 in sugar mill effluents.

The work also supported by Laxen *et al.* (1983) who studied the impact of effluent discharge from chloride–Industrial Batteries Ltd.

It is evident that the values of standard plate count (SPC) increased from December to April. The maximum average value recorded was 59.6×10^5/ml ± 7.51 in sugar mill effluents while minimum value noted was 49.0×10^5/ml ± 5.57 in IDPL wastes. It is concluded as the temperature increased SPC value rise up, BOD rise up but DO decreased.

According to Ghosh *et al.* (1976) at relatively higher temperature microbial activity increases and this results in increasing microbial population and BOD.

Study reveals that the value of most probable number (MPN) coliform bacteria, increases from December to April. The maximum average value noted was 36/100 ml ± 7.55 in sugar mill effluents.

The minimum average value recorded was 28/100 ml ± 7.00 in effluent of BHEL. MPN value rise up with increase in temperature.

So it is concluded that SPC, MPN and BOD proportional to temperature while DO inversely proportional to temperature. BOD value directly proportional to SPC and MPN while DO inversely propertional to SPC and MPN.

The effluents can not be considered suitable for drinking or bathing purposes, but can be disposed of for the irrigation.

Summary

The finding of this work conducted on the comparative microbial ecology of three different industrial effluents viz. IDPL, BHEL and Sugar mill duirng December, February and April in 1995–96. The text supported by 4 tables. IDPL situated in Rishikesh 5 km. away from city on Haridwar–Rishikesh road. BHEL situated in sector 4 and sector 5, 10 km. away from city in Haridwar. Sugar mill is situated near Railway line 2 km. away from city in Laksar.

Water samples were analysed by the standard methods.

The salient finding of the present study are given as below:

1. Water quality of three industrial effluents were investigated. The parameters determined were temperature, Total Solids, pH, Total Alkalinity, Total Hardness, Chlorides, Dissolved Oxygen (DO), Bio–chemical Oxygen Demand (BOD), Chemical Oxygen Demand (COD), free CO_2, Standard Plate Count (SPC), Most Probable Number (MPN).
2. The water temperature of the three industrial wastes was ranged from 16.3°C to 24.2°C.
3. In all the three industrial effluents total solids were observed in between 3156 mg/l to 9357 mg/l.
4. pH of different industrial wastes ranged from 6.5 to 7.8.
5. Total hardness of three industrial effluents varied from 210.0 mg/l to 370.6 mg/l in December and April respectively.
6. Total alkalinity of effluents of different industries recorded maximum was 214.7 mg/l in April and minimum 115.8 mg/l in December.

7. Chloride concentration of three industrial wastes ranged from 13.5 mg/l to 120.0 mg/l during December and April respectively.
8. Dissolved oxygen of three industrial effluents varied from December to April between 6.5 mg/l to 0.8 mg/l.
9. Biochemical oxygen Demand was observed maximum 405.5 mg/l during April and minimum 95.0 mg/l in December.
10. Chemical oxygen Demand in all the three industrial effluents was recorded minimum 120.3 mg/l and maximum 609.0 mg/l during December and April.
11. Free carbon dioxide of all industrial wastes was noted maximum 26.5 mg/l and minimum 7.5 mg/l during April and December.
12. Standard plate count of effluents of diferent industries ranged between maximum 67×10^5/ml and minimum 44×10^5/ml in February.
13. Most probable number of effluents was noted maximum 44/100 ml and minimum 21/100 ml during April and December.
14. High temperature since it promotes the microbial activities resulted in increased bacterial count and BOD. The DO obtained under thsese circumstances is obvious. Since oxygen is used in bio–chemical degradation of organic matter by microbes.
15. Total solids, BOD, COD, SPC, MPN values were always found relatively higher during April in comparison to December.
16. The water is not fit for drinking or bathing purposes. It can only be used for irrigation.

Chapter 2
Bathing Effect on Water Quality of Ganga

Introduction

In India, the water of many rivers has a unique place in all the religious activities. A large number of rivers and ponds have great religious significance. The religious sentiments of many pilgrims are so strong that they still regard these waters as pure and holy and drink it. Ceremonial bathing is also considered as a religious practice. For this reason several religious temples and Ashrams are located on the bank of these rivers and many of them have got constructed bathing ghats.

In nature, the quality of waters is ever changing because many constituents enter into natural bodies of waters through various activities of man. Man has been using them for disposal of liquid wastes from industries and communities. This results in increased degradation of their quality. A large number of pathogens, particularly bacteria and viruses are discharged from partially treated or un–treated sewage that pollute these waters. Their number decrease in the natural aquatic environments due to several physico–chemical and biological factors (*Narayanaswamy, 1982*).

Water has several beneficial uses as —

1. Drinking, bathing, swimming
2. boating, recreational
3. Source for public water supplies
4. Industrial and
5. Agricultural.

When waste water of different kinds enters a water source, its quality gets degraded and is no more suitable for the beneficial use. Water quality is expressed in terms of various physical (Turbidity, colour, odour, general appearance), chemical (all organic and inorganic chemical constituents) and biological (coliform organism, count, total plate count, species diversity indices) variables. The presence of excess concentration of these variables makes water unfit for a given beneficial use and that water is said to have become polluted for that particular use.

Due to rapid industrialization and various activities of man, a large volume of waste water of different kinds and characteristics is generated. Due to economic reasons all the generated waste water can not be treated and large volume of untreated waste waters are discharged into surface waters such as lakes, rivers and ponds which become polluted. The water thus polluted has great health hazard to the population and visitors who may consume such water in many ways. In older times, such danger was not significant as not much potentially dangerous waste water was disposed of into surface waters (*Bhargava, 1985*).

The river Ganga has a special place in Hindu mythology. It is closely inter–woven with our culture and civilization. It is the most important, significant pure and fast waste assimilative river of the Indo–Gangetic plain. The water of Ganga (Ganges) is considered the purest form of water and people carry it to different parts of the country and use it in various religious ceremonies as well as in other social practices.

The general direction of the river flow is from north–west to south–east. At its delta, the flow of generally southward. Ganga flows through one of the most fertile and densely populated territories in the world. Despite its religious importance, its length of 2,525 kilometres makes it one of the mightest rivers in the world. It originates from Gangotri in the Himalayan mountains at the height of approximately 25000 feet above sea level, and after flowing for about 1504 miles through agricultural land known as the Gangetic plain, falls into the Bay of Bengal near Kolkata. The important cities located on its bank are Rishikesh, Haridwar, Kanpur, Allahabad, Banaras, Patna and Kolkata. Its important tributaries are Yamuna, Gomti, Saraswati etc.

A large number of tributaries of varying dimensions merge their identity with this mighty river which is respected and worshipped by millions of people. Vast segment of Indian masses still regard the Ganga water to be the "purest form of liquid" in the world with the capacity to cure a wide range of human diseases. Even some of the highly educated and knowledgeable people believe that the Ganga is unpollutable. This type of blind faith is mainly responsible for greater and more severe ecological abuse of the Ganga–ecosystem. The north Indian states i.e. Uttar Pradesh, Bihar and Bengal are the main beneficiaries of the Ganga basin and in return they pollute its holy waters by discharging untreated effluents from tanneries, distilleries, sugar factories, paper mills and household activities. Human activities like bathing, washing, immersion of dead bodies and over fishing etc. add to this problem. (Bilgramy and Datta Munshi, 1985).

Although this river has the self purifying character as a result of which its quality recovers soon after it receives waste water, its pollution status at some places in its course, does reach a level that renders it less suitable for some of the beneficial uses (such as bathing, swimming and recreation, public water supplies after treatment, fish culture, agriculture and industrial). The Government of India has launched an ambitious project of monitoring the water quality of river Ganga with a view to clean it so that its self–purifying properties could be regained.

In India, numerous aspects of river pollution such as physico–chemical properties of water of different rivers (Ganapati and Chacko, 1951; Motwani *et al.*, 1956; Bhaskaran *et al.*, 1963; Deshmukh *et al.*, 1964; Mohan *et al.*, 1965; Mitra, 1982 and Raina *et al.*, 1984) and charges in biological composition of river with respect to the impact of pollutants (Chakraborty *et al.*, 1959, Venkateswarlu, 1966, 86; Rai, 1974; Bhatt *et al.*, 1985 and Nanden, 1985) have been studied. Water quality studies on the river Ganga have been the subject of study in recent years. These include the studies carried out between Gangotry and Rishikesh (Singh, 1986); Kanpur (Chattopadhya *et al.*, 1984 and Saxena *et al.*, 1966); at Varanasi (Agrawal *et al.*, 1976; Lakshminarayan, 1965; and Sikander and Tripathi, 1984); between Barauni to Farakka and Patna to Farakka, (Bilgrami and Datta Munshi 1979, 1985); between Katwa to Srirampore (Chaudhury, 1986) and between Bally to Bandel (Bagchi, 1986).

Systematic observations made at several points of Ganga where effluents are being regularly discharged reveal serious pollution problems (Roy and David, 1966; Ghosh *et al.*, 1973; Agrawal, *et al.*, 1976; Bhowmick and Singh, 1985; Sinha and Banerjee, 1987; and Shukla *et al.*, 1989). But only few studies have been made to find out the effect of bathing, burning of dead bodies on the water quality of river Ganga (Lakshminarayan, 1965; Agrawal *et al.*, 1976; Rai, 1978; Bilgrami and Datta Munshi, 1979; and Sikander, 1987).

As Haridwar is an important bathing place for the people of India, a large number of people from different parts of the country come over here to take their holy dip in the Ganga. The water of the Ganga popular as "Ganga Jal" is used by devotees and local people for drinking and to accomplish various religious customs. It is one of the seven great places of pilgrimage of Hindus in India. Great bathing ceremonies occur, especially on the first of Baisakh, the day on which Ganga first appeared on earth at the beginning of the Hindu solar year and on the occasion of Kumbh mela.

The present study was conducted to observe different physico–chemical parameters [temperature, pH, dissolved oxygen (DO), biological oxygen demand (BOD), alkalinity, acidity, chloride, calcium, hardness, total solids,] and micro–biological parameters [standard plate count, MPN (most probable number)] and isolation and observation of different fungal colonies cultured from the Ganga water in order to find out the effect of bathing on the quality of Ganga water at the three sampling sites (Har–ki–Pauri Ghat, Lakshman Jhoola Ghat, Prem Nagar Ashram Ghat) and their respective controls.

Results

The mean values (± SD) of different physico–chemical parameters like, temperature, pH, dissolved oxygen, biological oxygen demand, alkalinity, acidity, chloride, calcium, hardness, total solids and microbiological parameters like, standard plate count, and most probable number of coliforms observed at different sampling sites —

Site I (Har–ki–Pauri Ghat) and its control (Non–bathing place)

Site II (Lakshman Jhoola Ghat) and its control (Non–bathing place)

Site III (Prem Nagar Ashram Ghat) and its control (Non–bathing place) are shown in Tables 1, 2 and 3 respectively. Different types of fungi isolated fromm all the three sites and their control are shown separately.

1. **Temperature**

   There was not much variation in temperature of water at all three sites Har–ki–Pouri Ghat, Lakshman Jhoola Ghat, Prem Nagar Ashram Ghat and their controls. The minimum temperature (12.0°C) was observed at control of site I (Har–ki–Pouri Ghat) and maximum (12.85°C) at site II (Lakshman Jhoola Ghat). There was no significant difference in the temperature of the three sites, I, II, and III in comparison to their controls.

2. **pH**

   There was not much difference in the pH of water at the three sites Har–ki–pouri Ghat, Lakshman Jhool Ghat, Prem Nagar Ashram Ghat and their controls. However, minimum pH (7.2) was observed at control of site III (Prem Nagar Ashram Ghat) and maximum (7.8) at site I (Har–ki–Pauri Ghat), the pH of water at the three sites I, II and III was not significantly different for their controls.

3. **Dissolved Oxygen (DO)**

   The maximum value of DO was observed at the control of all the three sites. These were 10.5 mg/1 at the control of Prem Nagar Ashram Ghat, 10.2 mg/1 at the control of Lakshman Jhoola Ghat and 10.0 mg/1 at the control of Har–ki–Pauri Ghat. DO at all the three sites I, II and III was significantly lower with respect to their controls. It was 9.65 mg/1 at site I (Har–ki–Pauri Ghat), 9.41 mg/1 at site II (Lakshman Jhoola Ghat) and 9.35 mg/1 at site III (Prem Nagar Ashram Ghat).

4. **Biological Oxygen Demand (BOD)**

   The maximum BOD was observed at all the three sites (Har–ki–Pauri Ghat, 5.10 mg/1; Lakshman Jhoola Ghat, 4.75 mg/1 and Prem Nagar Ashram Ghat, 4.55 mg/1). The minimum value of BOD was found at the control of all the three sites (control of Har–ki–Pauri Ghat, 3.03 mg/1; control of Lakshman Jhool Ghat, 3.15 mg/1 and control of Prem Nagar Ashram Ghat, 3.59 mg/1). The BOD at the three sites was higher in comparison to their respective controls.

5. **Alkalinity**

The maximum value of alkalinity was observed at the three sites (Har–ki–Pouri Ghat, 72.4 mg/1; Lakshman Jhoola Ghat, 72.1 mg/1 and Prem Nagar Ashram Ghat, 70.12 mg/1). A minimum value of alkalinity was found at control of all the three sites (control of Har–ki–Pouri Ghat, 70.0 mg/1; Control of Lakshman Jhool Ghat, 69.5 mg/1 and control of Prem Nagar Ashram Ghat, 68.01 mg/1). The alkalinity at three sites were insignificantly higher than at their control places.

6. **Acidity**

The maximum acidity at the three sites was observed to be 5.75 mg/1 (Har–ki–Pouri Ghat), 5.35 mg/1 (Lakshman Jhoola Ghat) and 5.10 mg/1 (Prem Nagar Ashram Ghat). A minimum value of acidity was found at the control places of all the three sites (control of Har–ki–Pauri Ghat, 4.55 mg/1; control of Lakshman Jhoola Ghat, 4.15 mg/1 and control of Prem Nagar Ashram Ghat, 4.05 mg/1). The acidity at all the three sites was significantly higher than their respective control.

7. **Chloride**

The chloride was found to be maximum at all the three sites (Har–ki–Pauri Ghat 3.65 mg/1; Lakshman Jhoola Ghat, 3.32 mg/1 and Prem Nagar Ashram Ghat, 3.50 mg/1). It was minimum at control of all the three sites (control of Har–ki–Pauri Ghat, 2.45 mg/1, control of Lakshman Jhoola Ghat, 2.22 mg/1 and control of Prem Nagar Ashram Ghat, 2.15 mg/1). The chloride at all the three sites was significantly higher than their controls.

8. **Calcium**

The maximum calcium at all the three sites was observed to be 26.81 mg/1 at Har–ki–Pauri Ghat, 27.75 mg/1 at Lakshman Jhoola Ghat and 29.15 mg/1 at Prem Nagar Ashram Ghat. A minimum value of calcium was found at control of all the three sites (control of Har–ki–Pauri Ghat, 24.95 mg/1; control of Lakshman Jhoola Ghat, 26.08 mg/1 and control of Prem Nagar Ahsram Ghat, 28.12 mg/1). The calcium at the three sites was insignificantly higher with respect to their controls.

9. **Hardness**

The maximum hardness was observed at all the three sites. It was 94.6 mg/1 at Har–ki–Pauri Ghat, 81.5 mg/1 at Lakshman

Jhoola Ghat and 87.50 mg/1 Prem Nagar Ashram Ghat. The minimum hardness was found at control of the three sites. It was 85.50 mg/1 at the control of Har–ki–Pauri Ghat, 80.1 mg/1 at the control of Lakshman Jhoola Ghat and 85.75 mg/1 at the control of Prém Nagar Ashram Ghat. The value of hardness was significantly higher at all the three sites with respect to their controls.

10. Total Solids

The maximum total solids at all the three sites was observed to be 168.25 mg/1 at Har–ki–Pauri Ghat, 167.40 mg/1 at Lakshman Jhoola Ghat and 150.10 mg/1 at Prem Nagar Ashram Ghat. A minimum value of total solids was found at control of all the three sites (control of Har–ki–Pauri Ghat, 145.5 mg/1; control of Lakshman Jhoola Ghat, 155.5 mg/1 and control of Prem Nagar Ashram Ghat, 135.5 mg/1). Total solids were found to be significantly higher at all the three sites with respect to their controls.

11. Standard Plate Count (SPC)

The maximum SPC was observed at all the three sites. It was $13x10^3$/ml at Har–ki–Pauri Ghat, $12x10^3$/ml at Lakshman Jhoola Ghat and $10x10^3$/ml at Prem Nagar Ashram Ghat. The minimum SPC was found at control of all the three sites. It was $9x10^3$/ml at the control of Har–ki–Pauri Ghat, $8x10^3$/ml at the control Lakshman Jhoola Ghat and $6x10^3$/ml at the control of Prem Nagar Ashram Ghat. The value of SPC at all the three sites increased significantly in comparison to their controls.

12. Most Probable Number (MPN)

The maximum value of MPN was observed at all the three sites (290 coliforms/100 ml at Har–ki–Pauri Ghat, 240 coliforms/100 ml at Lakshman Jhoola Ghat and 160 coliforms/100 ml at Prem Nagar Ashram Ghat). The minimum value of MPN was observed at the control of all the three sites (240 coliforms/100 ml at Har–ki–Pauri Ghat, 210 coliforms/100 ml at the control of Lakshman Jhoola Ghat and 150 coliforms/100 ml at Prem Nagar Ashram Ghat).

13. Isolation of Fungi

A variety of fungi were cultured from the water of all the three sites and their controls.

Table 2.1: Physico–chemical parameter at site I (Har–ki–Pauri Ghat) and its control (Non bathing place)

(Mean Values ± S.D. of six observations from Nov. to Jan.)

Parameters	*Site I (Har–ki–Pauri Ghat)*			*Control (Non bathing place)*		
Temperature	12.50	±	1.12	12.00	±	1.15
pH	7.80	±	0.2	7.50	±	0.1
DO (mg/1)	9.65	±	0.15	10.00	±	2.5
BOD (mg/1)	5.10	±	0.15	3.00	±	0.25
Alkalinity (mg/1)	72.4	±	2.40	70.00	±	2.50
Acidity (mg/1)	5.75	±	0.13	4.55	±	1.5
Chloride (mg/1)	3.65	±	0.14	2.45	±	0.15
Calcium (mg/1)	26.81	±	1.50	24.95	±	1.42
Hardness (mg/1)	94.6	±	1.36	85.50	±	1.25
Total Solids (mg/1)	168.25	±	0.15	145.50	±	0.50
SPC* (No × 10^3/ml)	13.0	±	1.0	9.0	±	1.0
MPN* (Coliforms/100ml)	290			240		

* Microbiological Parameters

Table 2.2: Physico–chemical parameter at site II (Lakshman Jhoola Ghat) and its control (Non bathing place)

(Mean Values ± S.D. of six observations from Nov. to Jan.)

Parameters	*Site II (Lakshman Jhoola Ghat)*			*Control (Non bathing place)*		
Temperature	12.85	±	0.78	12.65	±	0.75
pH	7.50	±	0.2	7.40	±	0.1
DO (mg/1)	9.41	±	0.25	10.20	±	0.28
BOD (mg/1)	4.75	±	0.25	3.15	±	0.25
Alkalinity (mg/1)	72.1	±	2.51	69.50	±	2.33
Acidity (mg/1)	5.35	±	0.15	4.15	±	0.35
Chloride (mg/1)	3.32	±	0.25	2.22	±	0.50
Calcium (mg/1)	27.75	±	2.35	26.08	±	2.15

contd....

Table 2.2 – contd....

Parameters	Site II (Lakshman Jhoola Ghat)	Control (Non bathing place)
Hardness (mg/1)	81.5 ± 4.20	80.11 ± 4.40
Total Solids (mg/1)	167.40 ± 10.81	155.50 ± 11.75
SPC* (No × 10^3/ml)	12.0 ± 1.0	8.0 ± 1.0
MPN* (Coliforms/100ml)	240	210

* Microbiological Parameters

Table 2.3: Physico–chemical parameter at site III (Prem Nagar Ashram Ghat) and its control (Non bathing place)

(Mean Values ± S.D. of six observations from Nov. to Jan.)

Parameters	Site III (Prem Nagar Ashram Ghat)	Control (Non bathing place)
Temperature	12.75 ± 1.25	12.60 ± 1.50
pH	7.40 ± 0.2	7.20 ± 0.2
DO (mg/1)	9.35 ± 0.26	10.50 ± 0.21
BOD (mg/1)	4.55 ± 0.45	3.50 ± 0.50
Alkalinity (mg/1)	70.12 ± 1.50	68.01 ± 1.75
Acidity (mg/1)	5.10 ± 0.12	4.05 ± 0.15
Chloride (mg/1)	3.50 ± 0.15	2.15 ± 0.14
Calcium (mg/1)	29.15 ± 0.45	28.12 ± 0.50
Hardness (mg/1)	87.50 ± 5.40	85.75 ± 5.50
Total Solids (mg/1)	150.10 ± 7.80	135.50 ± 4.50
SPC* (No × 10^3/ml)	10.0 ± 1.0	6.0 ± 1.0
MPN* (Coliforms/100ml)	160	150

* Microbiological Parameters

Discussion

Pollution of natural surface waters is a common phenomenon. Rivers occupy an important religious place in India. They are traditionally considered sacred and bathing in these rivers is quite common during festivals. The religious significance of holy Ganga exceeds than that of any other river in the world. It has become a stream of agricultural, industrial, urban and human filth during its long journey as it collects surface run off and domestic garbage. To this is also added the burden of human activities like bathing, washing of clothes, immersion of ashes, unburnt corpses and dead animals.

The temperature of water controls the chemical reactions, the solubility of substances. It effects on the chemical and biological reactions of the organisms in water. The growth of nuisance of organism is enhanced by warm water condition and could lead to the development of unpleasant taste and odours (Sacramento, 1963). Mechalas *et al.* (1972) stated that man can function comfortably over long periods of time at a temperature range of 20 to 28.5°C. Higher temperature about 32°C causes undesirable physiological effects. The present study observed that there was not much variation in the temperature of water at all the three sites (bathing places) in comparison to their controls (Non bathing places). [World Health Organisation (WHO, 1971) and Indian Standards Institution (ISI, 1974) did not recommend any definite temperature value for suggesting potability and bathing of water respectively].

pH is the measure of the intensity of acidity or alkalinity and measures the concentration of hydrogen ions in water. In natural waters, pH also change diunnally and seasonally due to variation in photosynthetic activity which increases the pH due to consumption of CO_2 in the process. The microbiological integrity of water also depends upon its pH value (Bouwer, 1978). The present study showed that pH of all the three sites was not much different in comparison to their controls. The pH values were within desirable limit of WHO and permissible limit of ISI. Agarwal *et al.* (1976) also did not find much variation in pH at bathing ghats of Ganga river at Varanasi.

Dissolved oxygen is one of the most important parameter in water quality assessment and reflects the physical and biological processes prevailing in the water. Its deficiency directly affects the ecosystem of a river due to bioaccumulation and biomagnification. Maintenance and distribution of biota in aquatic ecosystems depends upon the concentration of DO to a great extent. High dissolved oxygen content is an indication of a healthy system (Bilgrami and Datta Munshi, 1985). The present study showed that DO in water of all the three sites was lower with respect to their respective controls. It was above the limit of DO recommended by WHO 1971 for supporting and maintaining healthy aquatic life. Besides this, the values were within permissible limit of ISI suggested for bathing. Similar values have been reported at Shiala and Raj ghats of river Ganga at Varanasi (Agarwal *et al.* 1976) and at Sangam during Ardha Kumbha in Allahabad (Upadhyay *et al.* 1982).

BOD is the amount of oxidizable organic matter present in the solution and the BOD value can be used as a measure of waste strength. The BOD test is useful in evaluating the self–purification capacities of streams which serves as a measure to assess the quantity of wastes which can be safely assimilated by the stream (cited in Trivedi and Goel, 1984). The present study observed that BOD at all the three sites was higher with respect to their controls. The BOD of water of the three sites were within standard limit recommended by WHO. During present study the values of BOD reported at bathing sites were almost similar to the values reported for the river Ganga at Bhandainai ghat in Varanasi (Agrawal *et al.*, 1976) and for the river Yamuna at Sangam during Ardha Kumbha (Upadhyay *et al.*, 1982).

The alkalinity of water is due to present of mineral salts present in it. It is primarily caused by the carbonate and bicarbonate ions. Alkalinity in itself is not harmful to human beings, still the water supplies with less than 100 mg/l are desirable for domestic use. The alkalinity makes water tasteful and helps in coagulation (Hussain, 1987). Acidity also depends upon the presence of decomposing vegetable matter and free carbon dioxide. In present study, alkalinity and acidity at all the three sites was higher than their respective controls. However WHO and ISI has not recommend any range of alkalinity/acidity for the potability of water and bathing respectively.

Chloride occurs naturally in all types of waters. Man and other animals excrete very high quantities of chloride together with nitrogenous compounds. Most of the water gets chloride in it when it flows through the area where salt is deposited (Gondave, 1985). In present study the chloride concentration at all the three sites was higher than their controls. However, the concentrations were within the highest desirable limit of WHO for drinking water and within the permissible limit of ISI for bathing. The values in present study were lower in comparison to values of chlorides at bathing ghat of Yamuna, Ganga at Sangam during Ardha Kumbha as reported by Upadhyay *et al.*, 1982.

Calcium is also one of the most abundant susbtances of the natural waters. Calcium, as such, has no hazardous effects on human health. In fact, it is one of the most important nutrients required by the organism. Concentration upto 1800 mg/l have been found not to impair any physiological reaction in man (Lehr *et al.*, 1980). Higher concentration of calcium is not desirable in washing, laundering and bathing owing to its suppression in formation of lather with soap (Trivedi and Goel, 1984). The present study observed that the calcium at all the three sites was higher with respect to their controls. WHO and ISI did not recommend any definite calcium value for its use as raw water for public water supply and bathing purpose.

The hardness of water is caused by bicarbonates, carbonates, sulphates, chlorides and nitrates of calcium and magnesium. Though hard water is not unfit for drinking but excessive hardness consumes more soap in laundries and forms deposits in boilers. Too soft water is tasteless (cited in Hussain, 1987). The increase in the hardness at bathing site can be attributed to the presence of more calcium in water.

On the basis of presence of concentration of calcium carbonate, Glohmann and Wasser (1976) classified the hardness of water in the following types :

6–75 mg/l	Soft
75–150 mg/l	Moderately hard
150–300 mg/l	Hard
300– up mg/l	Very hard

During present study water was observed to be 'moderately hard'. Besides this the hardness of water may also be due to metals such as iron, strontium and manganese (Sekerka and Lechnner, 1963). Hardness has no known adverse effects on health. However, some evidence has been given to indicate its role in heart disese (Peter, 1974). In present study hardness was observed to be slightly higher at all the three sites. But it was within limits set by WHO for potability of water.

The total solids in water comprise organic salts and small amount of organic matter. Generally, carbonates, bicarbonates, chlorides, nitrates and sulphates of sodium, potassium, calcium and magnesium contribute total solids in water. All these substances also influence the taste, hardness and corrosive property of water (Bruvold, 1970). Total solids at all the three sites were higher than their respective control sites. The total solids values obtained in the present study were slightly lower than those of the values reported at ghats of Varanasi (Agarwal *et al.*, 1976). The present study showed total solids were low in water at all the sites according to WHO standard.

The standard plate count (SPC) is an important parameter to know the density of aerobic and facultative bacteria in water sample which can grow at 37°C. The SPC values are useful in warning about excessive microbial growth in any water and also in judging the efficiency of water (cited in Trivedi *et al.*, 1987) Rai (1964, 78) reported that more the bacteria present, the greater will be the amount of organic matter which supports their nutritional requirements. The counts which are running to several thousands are definite indication of sewage pollution and a high count is sufficient to condemn water for drinking and bathing purposes. Ganapati (1957) and Rao *et al.*, (1979) have also stated that the number of bacteria developing is a measure of organic pollution. Bilgrami and Dutta Munshi (1985) also recorded many pathogenic bacteria in the bathing ghats at different centres. Bathers who drink untreated Ganga water are definitely open to risk to many intestinal disorders due to heavy concentration of coliform bacteria. In present study SPC was higher at all the three sites as compared to their controls. The higher values of SPC at these sites indicated more organic matter which is an indication of pollution.

In the present study the value of MPN was found to be higher at all the three sites at compared to their controls. The higher values of MPN at the sites indicate heavy sewage contamination. A variety of fungi were also isolated from the water of the three sites and their controls. The number of fungal colonies were more in case of all the three sites in comparison to their controls.

Bathing by a large number of people in a limited space at a particular ghat may transmit diseases through water which may be already contaminated by bathers themselves. Pathogenic bacteria in such conditions may produce diseases like intestinal disorders, dysentary, eye, nose, ear and throat infections, skin disorders etc.

Therefore, the persons suffering from contagious diseases should not be allowed to take bath.

Urinals and latrines need to be constructed on important busy ghats and nobody should be allowed to fulfil his nature calls in open. There should be separate arrangement for washerman's ghats and no washerman should be allowed to use any other ghat that specified for the purpose. Persons who come to take bath should not be allowed to wash their clothes using detergents at the bathing ghats. These detergents may cause serious pollution in water as they contain phosphates. The phosphates are responsible for the growth of algae which deplete the dissolved oxygen. There should be regular cleaning of bathing ghats to remove accumulated organic matter at the bottom.

The present study revealed that water of different sites viz., Site I, Site II and Site III can be used as a source of raw water for potable purposes after treatment; and for bathing and agriculture without treatment under the supervision of Public Health Authority. Futher, the present study also showed that Ganga which had been affected due to bathing, restored almost all its qualities at the control sites by the natural process of self–purification. It is suggested that more research is needed to find out the change, if any, in many other physico–chemical parameters and the types of bacteria present at the sites of all bathing places.

As Haridwar attracts a number of pilgrims, it needs close and continuous surveillance of Ganga river water quality by Public Health Authority to evaluate the health risks to bathers. A regular monitoring will not only prevent diseases and hazards but also check the water resources from going further polluted. General awareness and consciousness should be developed in public about ill effects of pollution through various sources like audio–visual aids, literature, posters etc.

Chapter 3
Water Quality of Ganga in Kumbh

Introduction

The water is one of the abundantly available substance in nature covering about 75% surface of earth's crust. It is an essential ingredient of all cellular organisms.

The basic source of water is precipitation. This is the water falling from the atmosphere to the surface of eath in the form of rain, snow etc. Out of the total estimates water in the earth and its atmosphere, only about 5 per cent of it is actually or potentially free and is in circulation and nearly 95 per cent of that is in the ocean. In contrast 95 per cent of earth water is bound in the lithosphere and in sedimentary rocks. The fresh water amount to only about 3 % of the total supply and three quarters of it is bound up in the polar ice caps in the Arctic and Antarctic oceans.

In India, all the major 14 rivers e.g. Ganga, Yamuna, Godavari, Gomti, Kosi, Cauvery, Ravi, Sone, Chenab, Jhelum, Narmada, Mani, Tapti and Krishna are held in high reverence by the people and are worshipped.

Haridwar is known for one of the holiest places of India. After every twleve year there is a big fair known as Kumbh. It is held at the banks of Ganges. Although there are several festivals those are celebrated that attract tourists at Haridwar all round the year, yet the Kumbh is supreme over all occasions. There are a large number of beautiful temples and ghats. That have added to the charm of this holy city.

Under the prevailing religious sentiments the water of the Ganges often referred to as 'gangajal' is regarded highly sacred.

The holy Ganges, the most sacred river of India, is one of the first ten mighty rivers of the world. With its tributaries, it is the largest and very important river basin of the country. It is a symbol of purity. The Alaknanda and the Bhagirathi are the two parent streams, which join at Devprayag to form the river Ganga. The two parent streams are snowfed and owe their origin to the Himalaya glaciers known as Gomukh. Ganga after traversing a distance of 2525 km. falls in to Ganga Sagar (West Bengal). The Ganges joins Ram Ganga at Kannauj and Yamuna at Allahabad.

No river evokes such feelings in the human mind as the Ganga does. She is loved not only by the Hindus but also non–Hindus including the non–religious people for one reason or another. And all this just because she is Ganga. The very name Ganga conjures up a picutre of holiness, of peace and joy of beauty and sweetness of all that is uplifting in the mind of a Hindu. Ganga is not just a river, she is a symbol of something higher than what the world has to offer. If the Himalaya is a Yogi steeped in meditation, Ganga arising from his heart is the culmination of that meditation. She is the bridge between heaven and earth.

In its 2525 km. long course Ganga dramatically varies her pace and form. Not more than a few feet wide in one Himalayas, it is tumbling boisterous stream in the hills and a deep fast flowing at Rishikesh. Ganga enters plains at Haridwar which is called 'Hari–ka–Dwar' by the Vaishnavities and Haridwar by the Shavities for the word 'Hari' means Vishnu and 'Har' Shiva and 'Dwar' means gate or to ways. From Haridwar the Ganga starts slowing down its pace.

Haridwar, the city is situated on the right bank of the Ganga while coming downwards. Not it is a district of Uttar Pradesh and it surrounded by Dehradun, Pauri and Bijnor districts. It is also surrounded by the Shivalik range so that it looks like a valley.

The river Ganga is the life line of millions of people. It is closely inter woven with our culture and civilisation. The health and well being of a large population is concerned with its water quality. It is the most important, significant pure and fast waste assimilative river of the Indogangetic plain. The general direction of the river flow is from north–west to south–east. It originates from Gangotri in the Himalayan mountains at the height of approximately 25,000 feet above sea level and after flowing for about 1,504 miles through agricultural land known as the Gangetic Plain, falls into the Bay of Bengal near Kolkata. The important cities located on its bank are Rishikesh, Haridwar, Kanpur, Allahabad, Banaras, Patna and Kolkata.

Today water resources have been the most exploited natural system since man strode the earth. Pollution of water bodies is increasing steadily due to rapid population growth, industrial proliferation, urbanization, increasing living standards and wide spheres of human activities. Chief polluting source of water are organic chemicals like detergents, pesticides, inorganic chemical such as metals, metal compound, salt and acid infectious agents from hospitals, slaughter houses, tanneries and large cities, plant nutrients like fertilizer, nitrate and phosphate.

Many big and small industrial units are established in Haridwar besides 180 Dharamshalas. The wastes released ultimately reach the Ganga. Domestic waste, effluent, garbage are almost directly discharged into the Ganga river. At this place the river also receives several 'nallah' like Bhimgoda nallah. The waste material and local effluents to the river ultimately change the physico–chemical proportion of the waters. During recent years, due to urbanisation and industrial growth. The quality of Ganga water has deteriorated considerably. During the past few decades a number of factories and hotels, have sprang up all along its route from the Himalaya and the Bay of Bengal and their poisonous waste products are polluting its water day to day. The cities are mainly responsible for the pollution of Ganga. Nearly whole of the city garbage and other waste material are thrown directly into Ganga river. There are more than 70 big industries situated along the side of Ganga. From these industries many tonnes of industrial wastes are released in the form of poisonous matter directly thus making this holy river to deadly poisonous. Ganga is now becoming dumping ground for poisonous chemicals from factories, agricultural wastes, garbages, acids and thousands of litres of sewage of various places.

It is estimated that about 1200 million litres of waste are released in Ganga from about 27 major cities situated at its banks. At Haridwar thousands of pilgrims take bath daily in the Ganga and degrades its water quality not only by bathing but also by dumping the things like ashes after last rites and garbage. The garbage is contributed by food items, packing material, urination, cotton and synthetic fibers, chemical detergents, medicines, glass, metallic containers and petroproducts. Many of the above items hare non–biodegradable. As a result, there is a heavy pollution load. Such activities are increased many fold on occasion like Baisakhi, Gurupurnima and Kumbh. At Varanasi, it receives the largest amount of biological and chemical filth from million of people taking bath every year.

At Kanpur, the main source of pollution of Ganga are jute, chemical, metal and surgical industries, tanneries, textile mills and bulk of domestic sewage of high organic nature.

Results

The results obtained from the observations made during November 1997 to March 1998, on the various physico-chemical and microbiological parameters are as under :

Sampling Station 'A' (Ganesh Ghat)

Physical Parameters

Table 3.1: Monthly variation in physico-chemical microbiological parameters of Ganga canal during November 1997 to March 1998 at sampling station 'A' (Ganesh Ghat of Ganga Canal).

Month	*Physical Parameters*				
	Air Temp. (°C)	*Water Temp. (°C)*	*Turbidity (JTU)*	*Transparency (cm)*	*TDS (mg/l)*
Nov.	22.0	14.0	60	78.2	81.5
Dec.	17.5	11.5	–	90.0	83.0
Jan.	11.0	10.0	–	96.0	70.0
Feb.	17.0	11.0	–	98.5	65.0
Mar.	19.0	13.5	–	94.6	81.0
Mean	17.3	12.0	12	91.46	76.1

1. **Air Temperature (°C)**
   The air temperature of bank of Ganga canal at this sampling station was observed to be the lowest in the month of January (11.0°C) while higher temperature was recorded in the month of November (22°C). The mean of air temperature at this sampling station was 17.3°C.

2. **Water Temperature (°C)**
   The water temperature of the Ganga canal at this sampling station was observed to be the lowest in the month of January (10.0°C) while the temperature recorded was higher in the month of November (14.0°C). The mean of water temperature at this sampling station was 12.0°C.

3. **Turbidity (JTU)**
   The turbidity was observed to be minimum (nil) during December to March at this sampling station. The maximum value was recorded 60 JTU in the month of November. The mean of turbidity was recorded 12 JTU.

4. **Transparency (CM)**
   The minimum value of transparency at this station was observed 78.2 cm in November and maximum value was recorded 98.5 cm in February. The mean as 91.46 cm.

5. **Total Dissolved Solid (mg/l)**
   The lowest value of TDS was observed (65.0 mg/l) in the month of February. The highest value was observed (81.5 mg/l) in November. The mean was recorded 76.1 mg/l.

Chemical Parameter

6. **pH**
   During the period of study the lowest value of pH was 7.6 in the month of March and highest value was 8.2 in the month of November. The mean of pH was recorded 7.94.

7. **Alkalinity (mg/l)**
   The Alkalinity in the Ganga canal at this sampling station was recorded to be lowest (81.0 mg/l) in the month of December and highest (90.6 mg/l) in the month of November. The mean of alkalinity at this sampling station was 85.36 mg/l,

Table 3.2 : Monthly variation in Physico-chemical microbiological parameters of Ganga canal during November 1997 to March 1998 at sampling station 'A' (Ganesh Ghat of Ganga Canal).

Month	*Chemical and Microbiological Parameters*								
	pH	*Alka-linity (mg/l)*	*Free CO_2 (mg/l)*	*DO (mg/l)*	*BOD (mg/l)*	*COD (mg/l)*	*Hard-ness*	*MPN (10^3/ml)*	*SPC (no./ml)*
Nov.	8.2	90.6	1.2	10.6	2.7	3.87	65.9	9.2	25.0
Dec.	8.0	81.0	0.97	11.3	2.4	3.43	63.0	9.2	19.0
Jan.	7.8	84.2	0.92	12.9	1.9	2.26	65.2	5.1	12.0
Feb.	8.1	88.6	0.89	13.0	1.6	1.62	68.5	9.2	29.0
Mar.	7.6	82.4	1.1	12.1	2.0	2.12	70.0	16.0	56.0
Mean	7.94	85.36	1.016	11.98	3.52	2.66	66.52	9.74	28.2

8. **Free Carbon Dioxide (mg/l)**
The lowest value of free carbon dioxide at this sampling station was recorded (0.89 mg/l) in the month of February. In the month of November, it was observed highest (1.2 mg/l). The mean was observed 1.016 mg/l.

9. **Dissolved Oxygen (mg/l)**
The value of dissolved oxygen of the Ganga canal at this sampling station was observed to be lowest (10.6 mg/l) in the month of November and highest (13.0 mg/l) in the month of February. The mean was recorded 11.98 mg/l.

10. **Bi-Chemical Oxygen Demand (BOD) (mg/l)**
The BOD of Ganga canal at this sampling station was recorded to be lowest (1.7 mg/l) in the month of February and higest (2.7 mg/l) in the month of November. The mean was recorded 3.52 mg/l.

11. **Chemical Oxygen Demand (COD) (mg/l)**
The COD of the Ganga canal at this sampling station was recorded to be lowest (1.62 mg/l) in February and highest (3.87 mg/l) in the month of November. The mean was recorded 2.66 mg/l.

12. **Hardness (mg/l)**
The hardness of water at this sampling station was observed

to be lowest in the month of December (63.0 mg/l). While higher in the month of March (70.0 mg/l). The mean of hardness at this sampling station was 66.82 mg/l.

13. **Standard Plate Count (SPC)**
The SPC of Ganga water at this sampling station was observed to be lowest in the month of January 12×10^3/ml and highest in the month of March 56×10^3/ml. The mean of SPC at this sampling station was 28.2×10^3/100 ml.

14. **Most Probable Number (MPN)**
The MPN of Ganga water at this sampling station was observed to be lowest in the month of January (5.1×10^2/100 ml) and highest in the month of March (16×10^2/100 ml). The mean of MPN at this sampling station observed 9.74×10^3/100 ml.

Sampling Station 'B' (Singh Dwar Ghat)

Physical Parameter

Table 3.3: Monthly variation in Physico-chemical microbiological parameters of Ganga canal during November 1997 to March 1998 at sampling station 'B' (Singh Dwar Ghat of Ganga Canal).

Month	*Physical Parameters*				
	Air Temp. (°C)	*Water Temp. (°C)*	*Turbidity (JTU)*	*Transparency (cm)*	*TDS (mg/l)*
Nov.	22.0	14.8	60	78.0	81.5
Dec.	16.5	12.5	–	90.5	88.0
Jan.	10.5	10.0	–	95.3	72.5
Feb.	17.5	12.5	–	98.2	64.2
Mar.	19.0	14.2	–	93.5	82.8
Mean	17.1	12.8	12	91.16	77.8

1. **Air Temperature (°C)**
The air temperature of bank of Ganga canal at this sampling station was observed to be the lowest in the month of January (10.5°C) while higher temperature was observed in the month of November (22.0°C). The mean of air temperature at this sampling station was 17.1°C.

2. **Water Temperature (°C)**
The water temperature of the Ganga canal at this sampling station was observed to be the lowest in the month of January (10.0°C) while higher in the month of November (14.8°C). The mean of water temperature at this sampling station was 12.8°C.

3. **Turbidity (JTU)**
The turbidity was observed to be minimum (nil) during December to March at this sampling station. The maximum value was recorded 60 JTU in the month of November. The mean of turbidity was 12 JTU.

4. **Transparency (cm)**
The minimum value of transparency at this station was observed 78.0 cm in November and maximum value was recorded in February 98.2 cm. The mean as 91.16 cm.

5. **Total Dissolved Solid (mg/l)**
The lowest value of TDS was observed in the month of February 64.2 mg/l. The highest value was observed in October 88.2 mg/l. The mean was recorded 77.8 mg/l.

Chemical Parameter

6. **pH**
During the period of study the lowest value of pH was 7.2 in the month of March and highest value was 8.5 in the month of November. The mean of pH was recorded 7.88.

7. **Alkalinity (mg/l)**
The Alkalinity in the Ganga canal at this sampling station was recorded to be lowest (81.6 mg/l) in the month of December and highest (96.0 mg/l) in the month of November. The mean of alkalinity at this sampling station was 83.84 mg/l.

8. **Free Carbon Dioxide (mg/l)**
The lowest value of free carbon dioxide at this sampling station was recorded (0.90 mg/l) in the month of February and highest (1.1 mg/l) in the month of November. The mean was observed 0.988 mg/l.

9. **Dissolved Oxygen (mg/l)**
The value of dissolved oxygen of the Ganga canal at this sampling station was observed to be lowest (10.7 mg/l) in the month of November and highest (13.4 mg/l) in the month of February. The mean of dissolved oxygen for the spot was 12.34 mg/l.

Table 3.4 : Monthly variation in Physico-chemical microbiological parameters of Ganga canal during November 1997 to March 1998 at sampling station 'A' (Singh Dwar Ghat of Ganga Canal).

Month	*Chemical and Microbiological Parameters*								
	pH	*Alka-linity (mg/l)*	*Free CO_2 (mg/l)*	*DO (mg/l)*	*BOD (mg/l)*	*COD (mg/l)*	*Hard-ness*	*MPN (10^3/ml)*	*SPC (no./ml)*
Nov.	8.6	86.0	1.1	10.7	2.9	3.93	64.8	29	9.2
Dec.	8.0	81.6	0.99	11.9	2.7	3.56	61.0	16	5.1
Jan.	7.6	83.7	0.95	12.6	2.0	2.67	62.5	10	2.2
Feb.	8.2	85.6	0.90	13.9	1.8	1.85	65.0	33	16.0
Mar.	7.5	82.3	1.0	12.6	1.9	1.92	69.8	51	16.0
Mean	7.98	83.84	0.988	12.34	2.26	2.786	64.62	27.8	9.7

10. Bio-Chemical Oxygen Demand (BOD) (mg/l)

The BOD of Ganga canal at this sampling station was recorded to be lowest in the month of February (1.8 mg/l) and highest in the month of November (2.9 mg/l). The mean was recorded 2.26 mg/l.

11. Chemical Oxygen Demand (COD) (mg/l)

The COD of the Ganga canal at this sampling station was recorded to be lowest (1.85 mg/l) in February and highest (3.93 mg/l) in the month of November. The mean of COD observed at this sport was 2.786 mg/l.

12. Hardness (mg/l)

The lowest value of hardness at this sampling station was recorded 61.3 mg/l in the month of December and highest in the month of March 69.8 mg/l. The mean of hardness of this sampling station was 64.62 mg/l.

13. Standard Plate Count (SPC)

The SPC of Ganga water at this sampling station was observed to be lowest in the month of January 10 × 10^3/ml and higher value was observed in the month of March 51 × 10^3/ml. The mean of SPC at this sampling station was 27.8.

14. Most Probable Number (MPN)

The MPN of Ganga water at this sampling station was observed to be lowest in the month of January (2.2 × 10^2/100 ml) and higher in the month of March ($16 \times 10^2/100$ ml). The mean of MPN at this sampling station observed 9.7.

Discussion

The pollution is the greatest crime of man kind against himself, with the rapid pace of industrialisation and many other sources. Water quality of Ganga river is getting degraded due to massive discharges of municipal waste and industrial waste of diverse origin.

The characteristics of Ganga canal water, studied presently included temperature, turbidity, transparency, TDS, pH, alkalinity, Free CO_2, Dissolved Oxygen, BOD, COD, Hardness, MPN, SPC. The sampling was done from November 1997 to March 1998.

The water temperature is one of the most important limiting factor for biological activities and internal metabolic reaction within the living organisms. The temperature of both of the stations was maximum in November and minimum in the month of January. A more or less similar trend has been observed in the stream Chakayadera in Garhwal by Dobriyal (1985) and Alkananda by Bodula and Single (1981).

The temperature should show an inverse relationship with the dissolved oxygen as reported by Das and dissolved oxygen is dissolved more during the period of active photosynthesis. Minimum dissolved oxygen was recorded in the month of November and maximum value of dissolved oxygen was observed in the month of February. This term is also stated by Khanna *et al.*, (1992) and Joshi (1996).

The BOD was observed maximum in November and minimum in February. A negative relationship has been observed between BOD and DO content. A similar pattern has been reported by Verma *et al.*, (1984) they noted during this period November 1997 to March 1998 temperature increase showed the increase in BOD. So it is stated that the temperature and DO shows inverse relationship. Bhargava (1985) also recorded higher BOD values during summer when the temperature was relatively higher. They attributed it to increase bacterial activity at relatively higher temperature.

Free carbon dioxide is released during the decomposition of certain susbstances and respiration of living organisms. Since higher temperature accelerates both the decomposition of organic susbtances as well as the respiratory activity of the biota. The direct relationship was established between the water temperature and free CO_2. Free CO_2 was observed maximum in the month of November and minimum in the month of February when the temperature of water decreases.

It is evident that the value of chemical oxygen demand (COD) was maximum in November and minimum in February. Accordingly to Bhatt *et al.*, (1988) and Joshi *et al.*, (1985) COD values do not differentiate between stable and unstable organic pollution, therefore, it is more than BOD values.

In the present study it was noted that the total solids were found maximum in the month of November and minimum in February. The total solid was closely inter–relation with one another and bear effect upon the aquatic life. Similar condition was also reported by Verma and Shukla (1969) and Khanna (1993) in their studies.

The higher transparency was recorded in February and lower in November. The same type of result was given by Khanna (1993).

In this study it was recorded that water was always slightly alkaline and various workers in their studies have shown similar results. Hardness is mainly due to the percentage of calcium or magnesium salt of bicarbonates, carbonates, sulphates and chlorides. The hardness is minimum in December and maximum in March. Similar result was obtained by Bhowmick and Singh (1985) and Bhatt *et al.*, (1985).

Alkalinity was due to presence of bicarbonates. The highest concentration was observed in November and lower in the month of December. Srivastava *et al.*, (1996) also reported similar trend in river Rapti.

It is evident that the values of standard plate count (SPC) increased from December to March. The maximum value was recorded in March and minimum value was recorded in January. It is concluded that as the temperature increased SPC value rise up, BOD rise up but DO decreased.

According to Ghosh *et al.*, (1976) at relatively higher temperature microbial activity increase and this results in increasing microbial population and BOD.

Study reveals that the value of most probable number (MPN) coliform bacteria, increases from December to March and maximum value noted in March and minimum in January.

So it is concluded that SPC, MPN and BOD are proportional to temperature while DO is inversely proportional to temperature, BOD value is directly proportional to SPC and MPN while DO inversely proportional to SPC and MPN.

Biological parameters have great importance for ecological point of river. All natural waters contain a variety of organisms both plants and animals besides plankton. Some alone were also found to be present. Kaushik and Ghankaksha (1988) also recorded the occurrence of the dormatophytes *Trichoderma koningi, T. bignorum,* besides on the fungi *Aspergillus candidus, A. flavous, A. terreus, A. niveus, Panicillium app.* and *Rhizopus sp.*

Chapter 4
Paper Mill's Effluents and Their Impact on Ganga Water

Introduction

Healthy environment is the most essential requisition of human life. It refers to natural things around us, which sustain life, such as earth atmosphere and healthy water. This water is one of the most important ingredients of plant and animal life. It has curious and unusual properties and plays a wonderful role in living system that is why "No life without water" is a common saying. It is a master solvent and all metabolic reactions of living organisms depend on the presence of water. Nearly three fourth of the earth's surface is covered by water, mainly ocean, and to a lesser degree rivers, lakes and streams but unfortunately today most of the surface and ground water has been polluted due to almost all kinds of human activities.

The defilement of water as a result of human activities is an old phenomenon. The ever increasing industrialisation, urbanisation and developmental activities and consequent pollution of water has brought a crisis of drinking water too. The water polllution is such contamination or such alteration of physico-chemical and biological properties of water stated by Sapru (1989). However, John (1977) defined, water pollution as the addition of an excess of materials or undesirable aquatic life to water or otherwise cause significant departure from the normal activities of various communities in or near bodies of water. According to Boumann (1965) pollution is caused due to solids, liquids and gases which are non-permissible, undesirable or objectionable.

These days the pollution of water is responsible for a very large number of mortalities and incapacitation in underdeveloped and developing countries. Most of the rivers of the world receive millions of litres of sewage, domestic waste, and industrial effluents containing substances varying in characteristic from simple nutrient to highly toxic substances.

It is noteworthy that industrial effluent and domestic waste is a major source of water pollution and about 70% water pollution is on account of industrial effluent, according to survey carried out by Central Pollution Control Board (CPCB). Both domestic and industrial waste constituents are the main pollutants to the water pollution. The industry continues to be one of the most significant cause of pollution of aquatic ecosystem due to diverse kind of waste produced by them. Polluted state of the water resource has led to a steady decline in fisheries and has also affected the irrigation of agricultural land. According to Quasim and Siddique (1969) in Indian rivers, however, it is the industrial effluent which is responsible for the major toxicity in water.

The water is an essential raw material in almost all manufacturing plants, though only a small part of it may appear in the final product. The remainder is the waste material and is likely to be contaminated to a small or large degrees.

Although there are many industires in India but among them the pulp and paper mill which has generated concern about the hazardous pollutant continuously released in water bodies is one of them. The continuous release of waste water into the water bodies not only cause aesthetic water pollution but also has been reported to be hazardous for aquatic flora and fauna (Nampoothery *et al.*, 1976; Roald, 1977 and Meleay, 1979).

In India about 288 pulp and paper mills manufacture 27.5 × 105 tonnes of paper per year (Gokhale *et al.* 1992) and manufacturing of paper yield large quantities of coloured and highly toxic waste water. It is estimated that about 273-355 m^3 of water is required per tonne of paper produced (Subramayam, 1975) that consequently generate 300 m^3 as waste water. According to Subramanayam and Hanumanulu (1976) the principal pollution in consequences of a pulp and paper mills are BOD, COD and suspended solids accompanied by the problems of colour, pH, temperature, dissolved oxygen and coliforms.

The volume and characteristic of waste water in pulp and paper mill depends on the type of manufacturing method adopted and the extent of reuse of water employed in Paper is a product manufactured in a continuous sheet by removing water from a slurry of cellulose fiber pulp. The pulp is produced by mechanically and or chemical processing of wood or other vegetative material to usable cellulosic fiber as an aqueous slurry.

The major portion of waste polluted water from making originate in the pulping process. About 80% of total pollution by paper mill is caused by pulp mill only (Srivastava *et al.*, 1990). In this process cell close raw material is digested under pressure and temperature mechanically a mixture of soda ash lime, magnesium calcium bisulphate and sulphuric acids in both alkaline sulphite process respectively. This digestion decomposed separates the binding non–cellosic materials such as lime and resins from the fiber. The ligning and nocellosic materials are dissolved leaving a stronger fiber for formation. The spent liquor from this process is known black liquor. The colour in this waste water coming out from the pulp mill section is due to lignin and lignin derivatives (Sastri, 1977).

After digestion chemically prepared wood pulp is blow into a closed pit where the black liquor is allowed to drain or to recovery process. The black liquor which is dark brown in colour constitute 10–15% of total waste water, but attributes 80% of the colour, 30% of the BOD and 60% of the COD of the total pollution load of pulp and paper mill effluent according to Subrahmanyan (1990).

In paper making process the pulp mixture is disintegrated and mixed in a beator to which are added various fillers and dyes to improve the quality of final paper product. Where, sometimes it is washed intially and produce a rather strong waste water, which is progressively diluted. This waste water is called white water, but according to Sicrra – Alvarezetal (1991). Thermochemically white water are non–toxic, containing readily biodegradable organic compounds.

For present study a typical large scale paper mill (Star Paper Mill Ltd.) located in North–Western Uttar Pradesh (Saharanpur) was selected. In this paper mill, the effluent is finally disposed of in the river Hindon, after about a 3 km. long running, whereas, it passes through residential area of mill and farming land. It is also used for irrigation purpose for agricultural lands.

The study reveals that the effluent quality of paper industry is relatively very poor while, it is compared with ISI values for both irrigation and disposal in river. The river water is so poor after confluence of effluents that it is neither fit for drinking nor for irrigation and it is also toxic to desirable aquatic life. So, effluent might be great threat to the inhabitants living on the bank of stream and river Hindon near the effluent discharge.

It is interesting to note the farsightedness of Vedic Scholars quoted by Kaushik (1988) that how some plants can help to get rid of microbes that are produced in polluted river. In the Atharvaveda some plants have been suggested to get rid of microbes that are produced in polluted river. It seems that we shall have to return to the Vedic fold to perpetuate our very survival with our highly polluted rivers.

Results and Discussion

The microbiological and physico–chemical characteristics of effluent and river water carried out from March to July 1995 from four stages.

Temperature of stage I, ranged from 21.5 to 34.2°C with an average value of 29.7°C; stage II, 19.3 to 31.5°C with an average value of 27.2°C, stage III, 17.0 to 28.2°C with an average value of 24.1°C; and stage IV, 18.5 to 30.2°C with an average value of 26.0°C. In all these cases it was observed that the temperature of all stages increases from March to July.

The pH of stage I, ranged from 7.4 to 8.6 with an average value of 8.1; stage II, 7.5 to 8.5 with an average value of 8.1; stage III, 7.1 to 7.8 with an average value of 7.5 and stage IV, 7.8 to 8.2 with an average value of 8.0 were determined. It indicated that pH value increases with temperature from March to July.

Total solids of stage I, ranged from 2454 to 2885 mg/l, with an average value of 2606 mg/l; stage II, 2527 to 2869 mg/l with an average value from 2723 mg/l; stage III, 315 to 338 mg/l with an average value of 329.3 mg/l; and stage IV, 831 to 882 mg/l with an average value of 852.6 mg/l.

Total dissolved solids of stage I, ranged from 1682 to 1785 mg/l with an average value of 1719 mg/l; stage II, 1384 to 1782 mg/l with an average value of 1586 mg/l; stage III, 181 to 187 mg/l with an average value of 184.3 mg/l; and stage IV, 503 to 549 mg/l with an average valueof 529.6 mg/l were estimated.

Total suspended solids of stage I, ranged from 873 to 1109 mg/l with an average value of 1028 mg/l; that of stage II, ranged from 946 to 1182 mg/l with an average value of 1071 mg/l; stage III, 139 to 159 mg/l with an average value of 149.6 mg/l; and stage IV, 344 to 384 mg/l with an average value of 358.3 mg/l.

Total hardness values of stage I, ranged from 2640 to 2883 mg/l with an average value of 2766 mg/l; that of stage II, 2580 to 2802 mg/l with an average value of 2688 mg/l; for stage III, 102 to 120 mg/l with an average value of 112 mg/l; and stage IV, 560 to 820 mg/l with an average value of 706.6 mg/l.

Calcium value of stage I, ranged from 1187 to 1346 mg/l with an average value of 1274 mg/l; that of stage II, 1052 to 1183 mg/l with an average value of 1131 mg/l; for stage III, 39.1 to 42.9 mg/l with an average value of 41.3 mg/l; and for stage IV, 195 to 220 mg/l with an average value of 209 mg/l.

Magnesium value of stage I, ranged from 755 to 865 mg/l with an average value of 812.3 mg/l; that of stage II, 750 to 856 mg/l with an average value of 806 mg/l; of stage III, 15.3 to 17.3 mg/l with an average value of 16.5 mg/l; and of stage IV, 96.2 to 103 mg/l with an average value of 100.6 mg/l.

Chloride values of stage I, ranged from 1065 to 1263 mg/l with an average value of 1183 mg/l; stage II, 1050 to 1249 mg/l with an average value of 1168 mg/l; stage III, 19.8 to 39.7 mg/l with an average value of 30.2 mg/l; and stage IV, 156 to 184 mg/l with an average value of 166.6 mg/l.

Carbon dioxide values of stage I, ranged from 4.4 to 8.8 mg/l with an average value of 7.3 mg/l; that of stage II, 6.6 to 13.2 mg/l with an average value of 10.2 mg/l; of stage III, 0 to 2.2 mg/l with an average value of 1.4 mg/l; and of stage IV, 4.4 to 6.6 mg/l with an average value of 5.8 mg/l.

Dissolved oxygen (DO) value of stage I was noted to range from zero to 1.5 mg/l with an average value of 0.5 mg/l. However, at stage II, the values of dissolved oxygen was nil. DO of stage III, ranged from 6.8 to 7.8 mg/l with an average value of 7.2 mg/l; and that of stage IV, 4.2 to 5.4 mg/l with an average value of 4.63 mg/l. It was observed that the DO value of all stages decreases from March to July.

BOD (Bio-Chemical Oxygen Demand) value of stage I, ranged from 88 to 180 mg/l with an average value of 110 mg/l; that of stage II, 165 to 305 mg/l with an average value of 250 mg/l; of stage III, 5 to 13 mg/l with an average value of 9.6 mg/l; and stage IV, 36 to 85 mg/l with an average value of 66.6 mg/l; showing increase from March to July.

COD (Chemical Oxygen Demand) value of stage I, ranged from 540 to 631 mg/l with an average value of 550.3 mg/l; that of stage II, 580 to 630 mg/l with an average value of 606.6 mg/l; of stage III, 13 to 21 mg/l with an average value of 17.8 mg/l; and of stage IV, 68 to 97 mg/l with an average value of 84.4 mg/l; were estimated.

The SPC (Standard Place Count) value in stage I, ranged from 15×10^4 to 45×10^4/ml with an average value of 30.3×10^4/ml; stage II, 31×10^4 to 57×10^4/ml with an average value of 46.3×10^4/ml; stage III, 4×10^4 to 13×10^4/ml with an average value of 9.3×10^4/ml; and stage IV, the values of SPC ranged from 15×10^4 to 39×10^4 with an average vaue of 25.3×10^4/ml were estimated. In all these cases SPC value increases from March to July.

The value of MPN (Most Probable Number) in stage I, ranged from 20 to 120/100 ml with an average value of 71.6/100 ml; that of stage II, 210 to 1100/100 ml with an average value of 523.3/100 ml; of stage III, 15 to 28/100 ml with an average value of 21/100 ml; and stage IV, 43 to 93/100ml and average value of 70.3/100 ml. It indicates that in all stages the value of MPN increases with increases in temperature. The BOD and the SPC show direct relationships.

Bhargava (1995) also reported high values of BOD, SPC and MPN when temperature was relatively higher. These findings are also supported by the work of Kamath (1985) when effluents were discharged into receiving waters at higher temperature. The DO values of such waters were found very low.

In microflora determination, the species of algae *Nitzschia palea Navicula viridural*, belong to Bacillariophyceae *Oscillatoria* sp. *Nostoc* sp. and *Microcystis* sp. to *Cyanophyceae*. The species of fungi isolated *Aspergillus niger, Aspergillus flavus* and *Penicillium Notatum* are indicative of their tolerance for such waters.

The presence of *Nitzschia palea, Navicula viridural, Microcystis* and *Oscillatoria* spp. are indicative of polluted water, as supported by Palmer (1969) who recorded that these genera are characteristic of polluted water. The species of *Cyanophyceae* were found to be more dominant in polluted water by Venketsverlu *et al.* (1987) too.

The present data indicate that the characteristics of stage I and II, included total solids, total dissolved solids, total suspended solids, total hardness, calcium and magnesium, and chloride concentration which vary from March to July. And in all cases, these are higher than Indian standard values for discharge of industrial effluent on land and in water bodies given in Table 3.5.

The BOD is an important indication of the level of biological pollution in waste water, as number of organisms increases the demand for oxygen increase proportionately. Depletion of oxygen in water lead to increase in BOD in the present data, (Tables 3.1, 3.2 and 3.4). It indicates that these effluents are highly polluted due to high range of BOD. These findings are also supported by the findings of Verma and Shukla (1969). The depletion of DO is responsible for polluted water was pointed out by Arora *et al.* (1974). Tiwari and Manzoorali (1989), Khann and Rajeekhar (1991) opined that the DO and the BOD are correlated to each other. These two simple parameters are correlated to each other to know the water quality. Therefore, these parameters can be used for the prediction of other parameters for monitoring water pollution.

The BOD and the COD values are recorded to be higher for the stage I and stage II, which are in consonance with the studies of Arokiasany *et al.* (1981) and these increase from March to July thereby indicating that the BOD and COD increase with the rise of temperature which is also supported by Balusu (1967), Singh *et al.* (1985) and Saxena *et al.* (1989).

Ghosh *et al.* (1976) expounded that at relatively higher temperature there is increased microbial activities causing the rise of BOD, SPC and MPN which is an asset to water quality. This supports the findings with regard to BOD, SPC and MPN recorded for the stage I, II and IV in this work. However, the BOD of the stage III, that of the river before confluence with the discharged effluent was recorded to be moderate. It is interesting to note that the DO of the stage III is maximum in comparison to effluents of the stages demarcated I, II and IV. But after confluence with the aforesaid effluents the BOD values are higher and values of DO were recorded to be significantly reduced as shown in Table 3.4 of the results. Jain *et al.* (1996) realised that the pollution of river due to discharge of waste effluents result into depletion of oxygen over a small stretch of river, which supports to the present data (Table 3.4). The findings by Ghatak and Konnar (1992), Paliwal (1983) and Singh *et al.* (1994) also support this work.

On confluence with effluents the river water turn from moderate hard to very hard. So, according to classification by Gloshman and Wasser (1976), high hardness and high chloride are indication of pollution. The carbon dioxide increases with decrease of DO level as stated by Ghatak and Kormar (1992) which is in concordance, with the present work carried out on the spoiling water of the river *Hindon.*

The results presented and the foregoing discussion reflects that the discharge of effluents in the river Hindon highly alters the water quality of this small river.

Summary

This dissertation presents a case study on pulp and paper mill's effluents and its impact, on river water quality in terms of microbiological and physico–chemical characteristics.

Here, four different stages were selected to carry on this study. These sampling stages were, (i), effluents at the outskirt of the mill, (ii), the stream of effluents followed by first stages used for irrigation, (iii), the river water before confluence and, (iv) the river water after confluence with the stream of industrial effluents.

The characteristic of effluents and river water which were studied included temperature, pH, total solids, total dissovled solids, total suspended solids, total hardness, calcium, magnesium, chloride, CO_2, DO, BOD, COD, SPC, MPN and algae and fungalmicro flora. In order to assess surface water pollution potential, the effluent characteristics were compared with ISI effluent standards.

The temperature of effluents at stages of outfall stream (I stage) and stream used for irrigation (II stage) was high. The addition of this effluents changed the temperature of down stream after confluence as compared to the up stream of preconfluence point but in all cases the observations were moderate in comparison to ISI standards and the pH of different stages varied with change of temperature. It was noted to be moderate while compared with ISI Standards.

The values of total solids, total dissolved solids and total suspended solids of effluent at stage of outfall stream and stream used for irrigation, were very high. These values were too high as compared to ISI standards. These higher values change the river water and enhanced aforesaid solids upto three times of down stream as compared to the up stream before confluence.

The values of hardness, calcium and magnesium of effluents were very high at outfall of the stream and the effluents stream part used for irrigation. Such type of effluents altered the down stream water quality and enhanced these values upto three times as compared with up stream before confluence.

Like hardness, the chloride were also noted to be high in the effluent stream at outfall and effluent stream part use for irrigation and similarly it enhanced the chloride value of the downstream while compared with the up stream waters of the river before confluence with effluent stream.

The value of carbon dioxide was seen to be highest in effluent stream used for irrigation, and its higher value affected the river water too, by enhancing the carbon dioxide value of down stream in comparison to the moderate up stream before confluences. It was observed that carbon dioxide value changed with temperature.

The biological characters DO and BOD were observed to be adversely related. The DO value of effluents was very low. While the values of BOD was high and the addition of the effluents reduced the DO value down stream river water, whereas the effluents adversely affected the value of BOD which got enhanced in comparison to up stream of river water before confluence with effluent.

The microbiological characters, MPN and SPC were high in case of effluent. These were dominant in effluent stream used for irrigation and these were noted to be enhanced in down stream of river water after confluence with effluents.

It was interesting note that the presence of the algae, *Nostoc* sp, *Oscillatoria* sp, *Microcysistis* sp, *Nitzschia palea, navicula viridural* from the down stream river after confluence, which were indicators of pollution according to Ghosh, *et al.* (1976) and the index of palmer (1969). The presence of fungi *Aspergillus niger, Aspergillus flavus* and *Penicillium spin* the river water after confluence with the effluents, indicated the addition of organic wastes.

Conclusion

The characteristic of effluent in terms of microbiological and physico–chemical parameters are very high with an exception to the DO concentration. In all the cases recorded values were very high in comparison to ISI values for both irrigation and disposal in big water bodies. After the entry of mill's effluents in the river Hindon, there is an increase in calcium, magnesium, chloride, total hardness, total dissolved solids, total suspended solids, pH, BOD, COD and CO_2. Dissolved oxygen was recorded in very low concentration as compared to up stream before confluence of the effluent, (stage III). At this stage, all the above said characteristics were observed in moderate condition. The reduction in DO concentration bear effect on the fishes.

This clearly indicates that the paper mills effluents are discharging wastage of in high concentration. Hence, it is altering the physico–chemical and biological characteristics of the river water significantly.

The present study indicates that certain members of algae in the water could be considered as indicators of pollution. However, further studies in this direction are needed to confirm the findings.

Chapter 5
Waste Water Effluents of Thermal Power Station

The waste water has double meaning — on the one hand, the waste water is water which usually has no value for enterprise; on the other hand this water carried harmful substances which contaminate water basins. From the stand point of an industrial enterprise (such as a power station), the term "waste water" can be applied to water that has been used in the technological process and for it poorer quality cannot be further used for the purpose. This water must be subjected to purification so as to remove the impurities from it at the plant (on in other techological process) or to discharge it into a water basin. The substances separated from waste water by purification can be either utilizable i.e. having a definite value (such as vanadium or nickel) or non–utilizable (refuse) i.e. having almost no value for economy (such as sand). The difference between the two is quite conditional since some substances which are now regarded as refuse might further become valuable when efficient methods for their utilization will be developed.

Thermal power stations are sources of the following types of waste water : (i) Cooling water which mainly cause thermal contamination of water basin; (ii) Waste water of water treatment plants and condensate cleaners; (iii) Waste water contaminated with petroleum products; (iv) Water used for washing of external surfaces of fuel oil fired stream generators and peak load heating boilers; (v) used solutions after chemical cleaning and preservation of thermal equipment; (vi) water from hydraulic ash disposal system of solid fuel fired thermal power stations; (vii) Municipal and domestic sewage waters; (viii) Used water after hydraulic cleaning of fuel conveying system; (ix) Rain (stream) water collected on the territory of a station.

Cooling waters of thermal power stations carry an enormous amount of heat into water basins. The effect of rejected heat on the ecological heat on the ecological systems of water basins, it is required by sanitary specifications that the temperature rise in water basins due to hot water discharge be not more than 5°C in winter time and 3°C in summer. Indeed, the temperature of water after condensers, even in recirculating cooling systems with cooling towers, does not exceed 26–30°C in winter and 35 to 42°C in summer. This water can be principally used for heating of green houses, fishing cribs and the like. But the utilization of hot water can only be seasonal nature and, besides, green houses, for instance require additional heating in winter time. Some of these methods can impair to some or other extent the quality of water in water basins. Water pollution by cooling system of thermal power stations occurs mainly due to petroleum products which pass into water from oil coolers of stream turbines.

Modern thermal power stations employ various schemes of water treatment depending on the quality of original water and requirements set to the make up water but all of them should generally include a water pretreatment plant ion-exchanger. The complete water pretreatment cycle includes co–agulation, treatment, magnesia desiliconization and filtering the stage of co–agulation and filtering being employed in all water pretreatment plants except for those where tap water is used. Waste waters of water pretreatment plants can in the general case contain unburned fuel, slime, coarse–disperse particles, organic substances, iron and aluminium compounds, $Mg(OH)_2$ and $CaCO_3$. The concentration of various impurities in waste water depends on the quality of the original water and the adopted methods of its pretreatments.

Waste water of condensate cleaners constitute only a small fraction of the total salt disposal from water treatment plants. They include water from loosening of sulpho–carbon filters, regeneration waste of iron exchanger etc. Waste water of water treatment plants contain various salts, acids and alkalis which have no specific toxic properties. These waste can however increase substantially the salts content in water basins and change the pH index. Waste water of water pretreatment plants also carry extracted organic substances which can rise the BOD of the basin, and certain suspended substances. For that reason, direct disposal of these waters into basins is not allowed. In particular, the sanitary specifications for water basins restrict the concentration of Cl ions to 350 mg/kg and that of SO_4^{2-} ions to 500 mg/kg. These ions are however present in waste waters of water treatment plants in substantially higher concentrations.

Petroleum products (various oils, sulphorous fuel oil, kerosene etc.) are especially dangerous for water basins. Petroleum products can enter water basins in emulsified, colloidal or dissolved state. The solubility of various petroleum products in wastes depends on their boiling point so that high boiling porudcts (such as fuel oils) are practically insoluble in water and their solubility can be neglected. Petroleum products can cause serious harm to warm basins by forming films on water surface and thus inhibiting natural aeration. On the other hand heavy products from bottom deposits thus isolating the bottom flora and fauna from the remaining portation of the basin. Petroleum products belong weakly oxidizable substances, so that their effect in water basin may be sensible for a long period. The final products of their oxidation are carbon dioxide and sulphur oxides. The main sources responsible for the appearance of petroleum products in waste waters of power stations are the fuel handling facilities, the main building, the electrical equipment (mainly transformers), and auxiliary services (such as locomotive sheds, garages and compressor houses). The loss of petroleum products at power stations should be mainly attributed to improper mounting to equipment negligent attitude of the personnel and also to emergency situations. At modern large power stations, the flow rate of petroleum contaminated waste waters does not usually exceed 100 m^3/h with oil concentration not more than 50 mg/kg.

Fuel oil ash can form deposits on the heating surfaces of boilers, especially in regeneration air heater where the deposition of ash may reach 10% of the total ash formed on the combustion of fuel oil. The product of fuel oil combustion contains suphur oxides SO_3 and SO_2. The sulphur deposits and the wet heating surface can easily catch fly ash from flue gases. Corrosion products and ash deposits increase the resistance of regenerative air heater for the passage of flue gases. For that reason periodic removal of deposits (usually in 15–20 day) is essential. This can be done by various methods, the main ones being as follows : (i) Washing with service water and alkaline solutions; (ii) Blowing with live or suspended steam; (iii) Flue gas calcination, with the air supply to the air heater cut off; (iv) Impulse methods; water washing can result in destruction of structural elements of the heater rotor and gas ducts. In washing of air heater, corrosion products, soot particles and ash deposits are washed off by water, their major portion being dissolved. Average concentration of impurities in wash water are presented below :

Impurities	*Concentration (g/kg)*
Coarse–disperse substances	0.5
Free H2SO4	4.0 – 5.0
Iron	7.0 – 8.0
Vanadium	0.3 – 0.8
Nickel	0.1 – 0.15
Copper	0.02 – 0.05
Dry residue	32.0 – 45.0

The process of the extraction of vanadium from fuel oil ash is based on the dissolution of ash by sulphuric acid, after which vanadium hydroxide is precipitated by means of ammonia in the presence of potassium choside. The precipitated V_2O_5 is dried, smelted and cast into the granules the containing 99.0% vanadium. Thus the most efficient method of vanadium extraction fuel oil ash comprises the following stages : smelting of ash with NaOH and KNO_3 – alkali zing in water – precipitation with H_2SO_4 solution washing – drying clacining (pelletizing).

The main equipment of a station (mainly boilers) is cleaned from scale and deposits by chemical washing. The procedure is called either prestarting or operating washing depending on whether a new or an operating boiler is being washed. The intervals between operating washings depend on the state of the equipment, usually one washing per year is sufficient. Operating washings are carried out in several stages which include water washing propes and treatment of heating surfaces with chemical reagents. The main disadvantage of this method is that the discharge of waste water increase of extensively varying rate, with the concentration and composition of impurities changing continuously. For that reason, the settling ponds for wash waters should have an ample capacity to receive the whole volume of wash water and dilute it three fold. In addition to the reagents used in washing procedures, waste waters contain in large amounts some harmful substances, for instance copper, zinc fluorine, hydrazine etc. Besides, waste waters of this type have a high BOD index i.e. much oxygen will be needed to oxidize their impurities in setting ponds.

Hydarulic ash disposal is based on hydraulic transport of ash. Ash is mixed with water to form pulp which is transported to ash sumps where coarse impurities settle down and the clarified water can be either discharges directly into water basin. The composition and content of impurities in water are determined by the chemical composition of ash, the system of ash removal, and on the availability of cleaning system of flue gases. Further, ash can contain minor quantity of toxic compounds of germanium, vanadium arsenic, mercury, berylium and fluorine which are of certain value for the national economy. Some carcinogenic substances can also be formed during fuel combustion and pass from ash to water. In addition, to certain components of ash water can also dissolve sulphur and nitrogen oxides and carbon dioxide when contacting with flue gases and with atmosphere. The most essential characteristics of clarified water from ash–disposal systems are the alkalinity, concentration of sulphates, total salt content, and concentration of toxic impurities. The pH index of clarified water in recirculating system of hydraulic ash removal may vary within a wide range, from strongly acid to strongly alkaline. Waste waters of hydarulic ash disposal can be discharged into water basins only when they contain no coarse–particle substances with the settling rate more than 0.25 to 0.75 mm/s. Besides, they must not change the pH index of the water basin beyond the specified limits 6.5 to 8.5 (Richter *et al.*, 1984).

In the present world of rapid industrialisation and modernisation, electricity is one of the cheapest, cleanest and safest mode of power but its production in coal–based power plant is not as clean because of its adverse impact on the environment. The development of nuclear power, coal–based power plants continue to have the maximum share of electric power generated in this country. The warm water effluents mainly affects the seasonal patterns of the limnological and hydrobiological properties of the receiving waters. Changes in the circulation conditons, increase in water temperature, decrease in the solubility of gases and changes in the receiving water's biotic communities are some of the effects which have been reported (Carins, 1956; Veiz and Ganon, 1960; Anderson, 1969, and Koschel and Mothes, 1976).

Heavy metals are toxic to most micro-organism at specific concentrations and often cause serious upsets in biological waste treatment plants. Heavy metals block the enzyme system or interfere with some essential matabolite of bacteria and protozoa (Morgan and Lackey, 1958). There are adverse effects of the metals presesnt in the fly ash on the vegetation grown on the fly ash disposal sites or the fly ash ponds. In addition to the traces of other metals, fly ash contain significant amounts of selenium, boron, copper, chromium, lead, molvedenum, aluminium and arsenic (Page *et al.*, 1978).

The accumulation of trace elements by terrestrial vegetation growing around ash disposal ponds or on fly ash landfil have been examined recently. Markedly elevated B levels were found in several harbaceous species near the ash disposal pond (Romney *et al.*, 1977). The values encountered would have been high enough to cause toxic symptoms in sensitive species, e.g. yellow sweet clover (*Melilokus officinalis*) found growing on land filled fly ash containing higher contents of trace elements as compared to controls (Furr *et al.*, 1976).

The possible contamination of water and soil by the leaching of the trace elements from the fly ash disposal ponds has only recently received attention. Theis *et al.*, (1978) have shown that leachates from these areas can contain significant levels of heavy metals, sulphates and total dissolved solids. Once a leachate containing various substances enters the soil environment, it is desirable to know the type and the soil particles i.e. the "natural attention capacity" of a specific soil component for a given leachate constituents. In the case of dissolved ions, these interactions will take the form of precipitation of a discrete soild phase or adsorption onto particles surfaces. They noted that the partitioning of traces elements between water and soil as well as their attenuation around the ash pond was affected in the pH and the oxides of maganese and iron. Theis and Richter (1979) reported that although trace element concentration in an ash pond were elevated levels rapidly decrease in ground water and soil distant from the pond. Accumulation of metals on the adjacent soil was primarily due to a complex interaction between precipitation and absorption on these metal oxides. It was pointed out by these workers that unless these metal oxides were replenished the absorptive capacity of the system was likely to be exhausted resulting in a free migration of traces elements from the ash pond. In view of its large absorptive capacity, maganese oxide addition to ash ponds may well be a useful control mechanism to reduce leaching of trace elements from such sites.

Factors affecting the element's chemical state and therefore its solubility behaviour with soil column, include soil type and its pH, the amount of rainfall, the acidity of the rainfall and the oxidation reduction conditions in the soil. The pH of ash pond leachate also exerts an important influence upon trace element behaviour. The alkaline nature of most fly ashes (Townsend and Gilham, 1975) results in the precipitation of insoluble hydroxides of many toxic elements in ash pond itself. Elements with anionic character, such as As, B, Cr, F, S, Mo and Se are more soluble in the alkaline environment and significantally elevated concentrations of these elements have been reported is ash ponds (Theis *et al.*, 1978) and their effluent water (Dreeson *et al.*, 1977; Turner *et al.*, 1982).

The direct run off fly ash leachate into adjacent streams has been found to result in the accumulation of several potentially toxic elements in both, invertebrate and vertebrate fauna and flora (Gutenamnn *et al.*, 1976; Cherry and Guthire, 1979; Cherry *et al.*, 1979). In an aquatic biomass study, Birge, (1978) showed that stimulated effluents from fly ash was particularly toxic to amphibians and fish. Moreover LC 50 values for several trace elements (As, Cd, Hg, Se and Zn) found in coal ash for these organisms were actually lower than the respective drainage water concentrations (Cherry *et al.*, 1979).

The thermal power plants in India produce about 35 to 40 million tonnes of fly ash every year and with increasing time more plants are coming up and this amount is bound to go up increasing the production of this wasteful menace. Keeping this in view present study was undertaken to find out the changes in various physico–

chemical properties viz. temperature, pH, total solids, alkalinity, carbon dioxide, hardness, calcium, magnesium, chloride, dissolved oxygen (DO), biochemical oxygen demand (BOD), chemical oxygen demand (COD) and microbiological properties viz. standard plate count (SPC) and most probable number (MPN) of microbes of waste water effluents being discharged from "Thermal Power Station (TPS), Bharat Heavy Electricals Limited (BHEL)", Ranipur, Haridwar, into the fields planted with Seesum (*Dalbergia sissoo*).

Results

The mean values (± SD) of different physico–chemical and microbiological parameters viz. temperature, pH, total solids, total dissolved solids, total suspended solids, alkalinity, free carbon dioxide, hardness, calcium, magnesium, chloride, dissolved oxygen, biochemical oxygen demand, chemical oxygen demand, standard plate count and most probable number at different sampling sites of thermal power station of BHEL, Ranipur, Haridwar viz., site I (control site), site II, site III, site IV and site V are depicted in tables 5.1 to 5.4. The percentage decrease/increase in various parameters at different sites II, III, IV and V are given in tables 2a–2d. The statistical analysis of results of physico–chemical and microbiological parameters observed at different sites are shown in tables 5.9–5.12.

Physico–Chemical Parameters

1. **Temperatre**

   There was significant ($P > 0.05$) difference in the temperature of water of TPS at different sites (II – V) in comparison to site I (control). The temperature varied from 10–12°C at site I, 25–30°C at site II, 22–25°C at site III, 16–20°C at site IV and 17–20°C at site V. The minimum value of temperature (11.0 ± 0.79) was observed at site I and maximum value of temperature (27.6 ± 2.51) was observed at site II (Tables 5.1, 5.9).

2. **Hydrogen Ion–concentration**

   There was no significant ($P < 0.05$) difference in the pH on different sites. It was found to be almost same at different sites (sites I – V) and it was in the range of 7.9 – 8.1, 8.1 – 8.2, 8.0 – 8.1, 8.0 – 8.1 and 8.0 – 8.1 at sites I, II, III, IV and V respectively (Tables 5.1, 5.9).

3. **Total Solid**

Total solids at site I were found to be the minimum (1.58 ± 0.20 ppm) at site II. The values of total solids at site III, site IV and site V increased significantly (P > 0.05) in comparison to the control site (site I). The values range from 1.4 – 1.8 ppm, 55.0 – 65.0 ppm, 5.0 – 10.0 ppm, 10.0 – 15.0 ppm and 15.0 – 20.0 ppm for the site I, II, III, IV and V respectively (Tables 5.1, 5.9).

4. **Total Dissolves Solids**

It ranged between 0.2 – 0.8 ppm, 40.0 – 42.0 ppm, 1.5 – 2.0 ppm, 1.0 – 2.0 ppm and 1.0 – 2.0 ppm for site I, II, III, IV and V respectively. The maximum value of TDS was found at site II (40.88 ± 1.04 ppm) and minimum value (0.56 ± 0.25 ppm) at site I. The value increased significantly (P > 0.05) at site III, IV and V in comparison to site I (Tables 5.1, 5.10).

5. **Total Suspended Solids**

TSS values ranged from 0.5 – 0.9 ppm, 18.2 – 19.0 ppm, 1.0–1.5 ppm, 0.5 – 0.7 ppm and 0.2 – 1.0 ppm for site I, II, III, IV and V respectively. The maximum value of TSS was found at site II (18.6 ± 0.32 ppm) and minimum value at site IV (0.52 ± 0.20 ppm). The value at site III increased insignicantly (P < 0.05) at site V in comparison to control site (Tables 5.1, 5.9).

6. **Alkalinity**

The minimum value of alkalinity 328.0 ± 21.68 ppm and maximum value 544.0 ± 37.82 ppm were observed at site IV and site V respectively. The alkalinity decreased insignificantly (P < 0.05) at site II and site III in comparison to site I (Tables 5.2, 5.10).

7. **Free Carbon Dioxide**

The range of free carbon dioxide exists between 8.8 – 26.4 ppm, 8.0 – 22.0 ppm, 17.4 – 35.2 ppm, 13.2 – 35.2 ppm and 17.4 – 35.2 ppm at site I, II, III, IV and V respectively. The minimum value 17.6 ± 4.69 ppm and maximum value 23.72 ± 7.40 ppm of free CO_2 were observed at site II and site V respectively. The value of CO_2 increased significantly (P > 0.05) at site III and site IV in comparison to the value of control site (site I) (Tables 5.2, 5.10).

8. **Hardness**

The minimum value of hardness at site II (215.33 ± 5.16) and maximum value were observed at site V (265.66 ± 4.97 ppm). There was a significant ($P > 0.05$) difference in the hardness at different sites II – IV but insignificant ($P < 0.05$) difference at site V in comparison to control site (site I). The values were found to be in the range of 200–240 ppm at site I, 210–220 ppm at site II, 220–230 ppm at site III, 210–246 at site IV and 260–270 ppm at site V (Tables 5.2, 5.10).

9. **Calcium**

The control site showed minimum calcium concentration (59.02 ± 3.04 ppm) and site V showed maximum concentration (71.64 ± 5.88 ppm). The calcium concentration was found to increase insignificantly ($P < 0.05$) at site II, site II, and significantly ($P > 0.05$) at site V (Tables 5.2, 5.10).

10. **Magnesium**

The value of magnesium was found to be maximum (21.05 ± 2.13 ppm) at site V and minimum (11.69 ± 4.00 ppm) was found to be at site IV. The values were insignificantly ($P < 0.05$) lower at site II and site IV and insignificantly ($P < 0.05$) and significantly ($P > 0.05$) higher at site III and site V respectively in comparison to site I (Tables 5.2, 5.10).

11. **Chloride**

There was a maximum (32.94 ± 2.33 ppm) and minimum (14.73 ± 0.82 ppm) concentration of chloride at site I and site II respectively. The chloride concentration decreased significantly ($P > 0.05$) at site II, at site III, at site IV and site V in comparison to control site. (Tables 5.3, 5.11).

12. **Dissolved Oxygen (DO)**

The value of DO was found to be maximum (15.89 ± 0.44 ppm) at site V and minimum (1.86 ± 0.89 ppm) at site II. The value of DO decreased significantly ($P > 0.05$) at site II, site III and site IV but significantly ($P > 0.05$) increased at site V in comparison to the control site (Tables 5.3, 5.11).

13. **Biochemical Oxygen Demand (BOD)**

The minimum value of BOD was found at site II (0.97 ± 0.22 ppm) and maximum value at site V (9.67 ± 0.88 ppm). The values of BOD increased significantly ($P > 0.05$) at site III, at site IV and site V, but decreased significantly ($P > 0.05$) at site II in comparison to site I (Tables 5.3, 5.11).

Table 5.1: Physico–chemical parameters at different sites of TPS at BHEL Ranipur, Haridwar. (Values are mean ± SD of five observations each)

Parameters	*Site I*	*Site II*	*Site III*	*Site IV*	*Site V*
Temperature	11.0 ± 0.79	27.6 ± 2.51	23.4 ± 1.34	18.2 ± 1.79	18.4 ± 1.52
(° C)	(10.0 – 12.0)	(25.0 – 30.0)	(22.0 – 25.0)	(16.0 – 20.0)	(17.0 – 20.0)
pH	8.0 ± 0.20	8.16 ± 0.49	8.02 ± 0.21	8.04 ± 0.41	8.04 ± 0.41
	(7.9 – 8.1)	(8.1 – 8.2)	(8.0 – 8.1)	(8.0 – 8.1)	(8.0 – 8.1)
TS	1.58 ± 0.20	61.8 ± 5.49	7.6 ± 2.51	12.4 ± 2.51	13.2 ± 5.2
(ppm)	(1.40 – 1.80)	(55.0 – 65.0)	(5.0 – 10.0)	(10.0 – 15.0)	(15.0 – 20.0)
TDS	0.56 ± 0.25	40.88 ± 1.04	1.9 ± 0.42	1.4 ± 0.39	1.4 ± 0.39
(ppm)	(0.2 – 0.8)	(40.0 – 42.0)	(1.5 – 2.5)	(1.0 – 2.0)	(1.0 – 2.0)
TSS	0.72 ± 0.20	18.6 ± 0.32	1.22 ± 0.26	0.52 ± 0.20	0.6 ± 0.37
(ppm)	(0.5 – 0.9)	(18.2 – 19.0)	(1.0 – 1.5)	(0.2 – 0.7)	(0.2 – 1.0)

Range in parentheses

Table 5.2: Physico–chemical parameters at different sites of TPS at BHEL Ranipur, Haridwar. (Values are mean ± SD of five observations each)

Parameters	*Site I*	*Site II*	*Site III*	*Site IV*	*Site V*
Alakalinity	380.0 ± 30.82	370.0 ± 57.0	360.0 ± 41.83	328.0 ± 21.68	544.4 ± 37.82
(ppm)	(350.0 – 420.0)	(300.0 – 450.0)	(300.0 – 400.0)	(300.0 – 350.0)	(500.0 – 600.0)
Free CO_2	18.48 ± 7.24	17.6 ± 4.69	22.88 ± 7.23	22.88 ± 8.46	23.72 ± 7.40
(ppm)	(8.8 – 26.4)	(8.8 – 22.0)	(17.6 – 35.2)	(13.2 – 35.2)	(17.4 – 35.2)
Hardnesss	221.6 ± 20.41	215.33 ± 5.16	225.33 ± 4.50	231.0 ± 4.69	265.66 ± 4.97
(ppm)	(200.0 – 240.0)	(210.0 – 220.0)	(220.0 – 230.0)	(210.0 – 246.0)	(260.0 – 270.0)
Calcium	59.02 ± 3.04	60.19 ± 8.77	63.49 ± 3.65	68.62 ± 5.24	71.64 ± 5.88
(ppm)	(55.31 – 62.52)	(58.52 – 62.52)	(60.12 – 68.94)	(60.12 – 72.14)	(65.73 – 76.15)
Magnesium	17.23 ± 2.95	16.96 ± 0.8	17.97 ± 0.80	11.69 ± 4.00	21.05 ± 2.13
(ppm)	(15.08 – 20.47)	(16.08 – 17.54)	(17.06 – 18.52)	(8.77 – 16.08)	(19.49 – 23.39)

Range in parentheses

Table 5.3: Physico–chemical parameters at different sites of TPS at BHEL Ranipur, Haridwar. (Values are mean ± SD of five observations each)

Parameters	*Site I*	*Site II*	*Site III*	*Site IV*	*Site V*
Chloride (ppm)	32.94 ± 2.33 (31.24 – 35.50)	19.59 ± 1.56 (18.46 – 21.30)	14.73 ± 0.82 (14.20 – 15.62)	18.18 ± 3.24 (15.62 – 21.30)	19.31 ± 1.61 (17.04 – 21.30)
DO (ppm)	11.59 ± 1.13 (10.14 – 12.17)	1.86 ± 0.89 (1.21 – 2.84)	3.41 ± 0.49 (3.24 – 3.65)	7.62 ± 0.44 (7.29 – 8.11)	15.89 ± 0.44 (15.41 – 16.22)
BOD (ppm)	4.38 ± 0.44 (4.05 – 4.86)	0.97 ± 0.22 (0.81 – 1.22)	4.54 ± 0.67 (4.05 – 5.27)	6.65 ± 0.22 (6.49 – 6.89)	9.67 ± 0.88 (8.11 – 10.14)
COD (ppm)	4.16 ± 1.80 (2.4 – 6.0)	33.20 ± 0.86 (32.0 – 34.0)	14.52 ± 2.62 (13.0 – 16.0)	11.68 ± 1.70 (10.0 – 14.0)	9.20 ± 0.78 (8.0 – 10.0)

Range in parentheses

Table 5.4: Microbiological parameters at different sites of TPS at BHEL Ranipur, Haridwar. (Values are mean ± SD of five observations each)

Parameters	*Site I*	*Site II*	*Site III*	*Site IV*	*Site V*
SPC (10^4/ml)	17.50 ± 2.50 (15.0 – 20.0)	13.60 ± 2.93 (10.0 – 18.0)	54.0 ± 4.18 (50.0 – 60.0)	79.0 ± 4.18 (75.0 – 85.0)	87.0 ± 5.36 (80.0 – 95.0)
MPN (per 100 ml)	0.36 ± 0.03 (0.3 – 0.4)	1.80 ± 0.62 (0.7 – 2.1)	2.74 ± 1.68 (0.3 – 4.3)	8.20 ± 6.64 (1.5 – 15.0)	21.16 ± 15.75 (2.8 – 46.0)

Range in parentheses

14. Chemical Oxygen Demand (COD)

The control site showed minimum value (4.16 ± 1.80 ppm) of COD and site II showed maximum value (33.20 ± 0.86 ppm). The value of COD increased insignificantly ($P < 0.05$) at site V but significantly ($P > 0.05$) increased at sites II and IV in comparison to site I (Tables 5.3, 5.11).

Microbiological Parameters

1. Standard Plate Count (SPC)

The value of SPC was found to be maximum (87.0 ± 5.36 ppm) at site V and minimum (13.6 ± 2.93 ppm) at site II. The value of SPC was significantly ($P > 0.05$) lower at site II but significantly ($P > 0.05$) higher at sites III, IV and V in comparison to control site (Tables 5.4, 5.12).

2. Most Probable Number (MPN)

The minimum value was found to be site I (0.36 ± 0.03/ 100 ml) and maximum value at site V (21.16 ± 15.75/100 ml) of MPN. The MPN values increased significantly ($P > 0.05$) at sites II, III, IV and V in comparison to control site (Tables 5.4, 5.12).

Discussion

A study on the nature of water and waste water effluents of thermal power station is essential in the design and operation of thermal power station. The efficiency of TPS is based on the basic of coal fired and flow rate of water. The processing brings changes in physico–chemical and microbiological properties of raw water entering into TPS and results in the production of fly ash and warm waste water effluents.

Various physico–chemical parameters like termpature, pH, total solids, TDS, TSS, alkalinity, free CO_2, hardness, calcium, magnesium, chloride, dissolved oxygen, BOD, COD and microbiological parameters like bacterial count have a significant role in determining the pollution load of waste water effluents (Thomas, 1953; Sacramento, 1963; Webber and Stumn, 1963; APHA, 1976; Lehr *et al.*, 1980; Richter *et al.*, 1984; Trivedy and Goel, 1984).

The changes in the temperature, solubility of gases and water's biotic communities of the pond/ground wates due to the waste water effluents of TPS have been reported (Velz and Gannon, 1960;

Table 5.5: Percentage decrease/increase in different physico-chemical parameters at different sites of TPS at BHEL Ranipur, Haridwar. (Values are mean ± SD of five observations each).

Parameters	*Site II*	*Site III*	*Site IV*	*Site V*
Temperature	+ 150. 909	+ 112.729	+ 65.454	+ 67.272
pH	+ 2.0	+ 0.25	+ 0.5	+ 0.5
TS	+ 3811.392	+ 381.012	+ 684.810	+ 735.443
TDS	+ 7200.0	+ 239.285	+ 150.0	+ 150.0
TSS	+ 2483.333	+ 69.444	+ 27.777	+ 16.666

Percentage decrease/increase calculated from the control taking it as centpercent.

Table 5.6: Percentage decrease/increase in different physico-chemical parameters at different sites of TPS at BHEL Ranipur, Haridwar. (Values are mean ± SD of five observations each).

Parameters	*Site II*	*Site III*	*Site IV*	*Site V*
Alkalinity	– 2.631	– 3.157	– 13.684	+ 43.157
Free CO_2	– 4.761	+ 23.809	+ 23.809	+ 28.354
Hardness	– 2.855	+ 1.655	+ 4.213	+ 19.850
Calcium	+ 2.002	+ 7.574	+ 16.266	+ 21.394
Magnesium	– 1.567	+ 4.283	– 32.153	+ 22.170

Percentage decrease/increase calculated from the control taking it as centpercent.

Table 5.7: Percentage decrease/increase in different physico-chemical parameters at different sites of TPS at BHEL Ranipur, Haridwar. (Values are mean ± SD of five observations each).

Parameters	*Site II*	*Site III*	*Site IV*	*Site V*
Chloride	– 40.517	– 55.293	– 44.827	+ 41.379
DO	– 83.939	– 70.631	– 34.267	+ 37.062
BOD	– 77.798	+ 3.70	+ 51.850	+ 120.831
COD	+ 689.076	+ 299.038	+ 180.769	+ 121.153

Percentage decrease/increase calculated from the control taking it as centpercent.

Table 5.8: Percentage decrease/increase in different microbiological parameters at different sites of TPS at BHEL Ranipur, Haridwar. (Values are mean ± SD of five observations each).

Parameters	*Site II*	*Site III*	*Site IV*	*Site V*
SPC	– 22.285	+ 208.571	+ 351.428	+ 397.142
MPN	+ 400.0	+ 601.111	+ 2177.777	+ 5777.777

Percentage decrease/increase calculated from the control taking it as centpercent.

Table 5.9: Statistical analysis of results of physico–chemical parameters observed at different sampling sites of TPS of BHEL Ranipur, Haridwar.

Sites	*"t" values*				
	Temperature	*pH*	*TS*	*TDS*	*TSS*
Site I (control) vs Site II	14.127 ($P > 0.05$)	0.617 ($P < 0.05$)	24.499 ($P > 0.05$)	85.063 ($P > 0.05$)	106.428 ($P > 0.05$)
Site I (control) vs Site III	17.867 ($P > 0.05$)	0.675 ($P < 0.05$)	5.351 ($P > 0.05$)	6.203 ($P > 0.05$)	3.448 ($P > 0.05$)
Site I (control) vs Site IV	8.256 ($P > 0.05$)	0.425 ($P < 0.05$)	9.617 ($P > 0.05$)	4.117 ($P > 0.05$)	1.574 ($P < 0.05$)
Site I (control) vs Site V	9.711 ($P > 0.05$)	0.425 ($P < 0.05$)	4.991 ($P > 0.05$)	4.117 ($P > 0.05$)	0.227 ($P < 0.05$)

Significant difference given in parantheses.

Zawisza and Backiel, 1972; Koschel and Mothes, 1976; Eloranta, 1983; Richter *et al.*, 1984; Simsiman *et al.*, 1987; MPPCB, 1994). However, there seems to be no report on the characteristics of physico–chemical and microbiological parameters of waste–water effluents of TPS.

During present study the physico–chemical characteristics of TPS at site I, entering in to the TPS showed a temperature of 11°C, pH of 8.0, total solids of 1.58 ppm, TDS of 0.56 ppm, TSS of 0.72 ppm, alkalinity of 380.0 ppm, free CO_2 of 18.48 ppm, hardness of 221.66 ppm, Ca of 59.02 ppm, Mg of 17.23 ppm, Cl of 32.94 ppm,

Table 5.10: Statistical analysis of results of physico–chemical parameters observed at different sampling sites of TPS of BHEL Ranipur, Haridwar.

Sites	*"t" values*				
	Alkalinity	*Free CO_2*	*Hardness*	*Calcium*	*Magnesium*
Site I (control) vs Site II	0.345 (P < 0.05)	0.318 (P < 0.05)	0.736 (P < 0.05)	0.284 (P < 0.05)	0.199 (P < 0.05)
Site I (control) vs Site III	0.860 (P < 0.05)	1.959 (P < 0.05)	0.430 (P < 0.05)	2.108 (P < 0.05)	0.286 (P < 0.05)
Site I (control) vs Site IV	3.085 (P > 0.05)	1.05 (P < 0.05)	1.092 (P < 0.05)	3.043 (P < 0.05)	2.248 (P < 0.05)
Site I (control) vs Site V	7.517 (P > 0.05)	1.39 (P < 0.05)	5.129 (P > 0.05)	4.248 (P > 0.05)	2.345 (P > 0.05)

Significant difference given in parantheses.

Table 5.11: Statistical analysis of results of physico–chemical parameters observed at different sampling sites of TPS of BHEL Ranipur, Haridwar.

Sites	*"t" values*			
	Chloride	*DO*	*BOD*	*COD*
Site I (control) vs Site II	10.652 (P > 0.05)	14.145 (P > 0.05)	15.768 (P > 0.05)	32.665 (P > 0.05)
Site I (control) vs Site III	16.485 (P > 0.05)	23.002 (P > 0.05)	25.121 (P > 0.05)	10.257 (P > 0.05)
Site I (control) vs Site IV	14.159 (P > 0.05)	11.616 (P > 0.05)	10.590 (P > 0.05)	7.272 (P > 0.05)
Site I (control) vs Site V (P > 0.05)	10.767 (P > 0.05)	12.564 (P < 0.05)	12.077 (P > 0.05)	0.515 (P < 0.05)

Significant difference given in parantheses.

Table 5.12: Statistical analysis of results of microbiological parameters observed at different sampling sites of TPS of BHEL Ranipur, Haridwar.

Sites	*"t" values*	
	SPC	*MPN*
Site I (control) vs Site II	2.321 (P > 0.05)	5.161 (P > 0.05)
Site I (control) vs Site III	16.689 (P > 0.05)	3.160 (P > 0.05)
Site I (control) vs Site IV	28.152 (P > 0.05)	2.640 (P > 0.05)
Site I (control) vs Site V	26.218 (P > 0.05)	2.952 (P > 0.05)

Significant difference given in parantheses.

dissolved oxygen of 11.59 ppm, BOD of 4.38 ppm and microbiololgical characteristics i.e. SPC of 17.5×10^4/ml and MPN of 0.36/100 ml (Tables 5.1–5.4).

At site II (Before ash tank), the waste water effluents of TPS showed that the temperature (27.6°C) was significantly (P > 0.05) very high in comparison to control site (11.0°C). There was a significantly (P > 0.05) increase in temperature (150.909%), total solids (3811.392%), TDS (7200.00%), TSS (2483.333%), COD (698.076%) and MPN (400.00%). But there was insignificant (P < 0.05) increase in pH (2.0%) and Ca (2.002%). Besides this the insignificant (P < 0.05) decrease was observed in alkalinity (2.631%), free CO_2 (4.761%), hardness (2.855%) and Mg (1.567%) and significant (P > 0.05) decrease in case of Cl (40.517%), disolved oxygen (83.939%), BOD (77.789%) and SPC (22.285%), (Tables 5.5–5.12). There was maximum change in the values of different parameters like temperature, total solids, TDS, TSS ,Cl, DO, BOD, COD, SPC and MPN due to the containing of high temperature and maximum solids in waste water effluents from the furnace resulting from the coal burning into the furnace.

At site III (after ash tank and pollution tank), the waste water of TPS at this site showed percentage decrease/increase in various parameters. There was insignificant ($P < 0.05$) decrease in values of alkalinity (3.157%) but significant ($P > 0.05$) decrease in DO (70.63%) and in Cl (55.293%) in comparison to the parameters of control site. Beside this there was insignificant ($P < 0.05$) increase in pH (0.25%), free CO_2 (23.809%), hardness (1.655%), Ca (7.574%) and Mg (4.283%) and the significant ($P > 0.05$) increase in the temperature (112.729%), total solids (381.012%), TDS (239.285%), TSS (69.444%), BOD (3.700%), COD (249.038%), SPC (208.571%) and MPN (601.111%), (Tables 5.5–5.12). At this site the values of all parameters (except i.e. alkalinity, dissolved oxygen and chloride) were higher in comparsion to parameter of site I. The decrease/increase in values of all the parameters of this site was due to the treatment of waste water effluents into the ash tank and water pollution tank. In these tanks the maximum solids settle down and few very fine particles are left. The ash tank provided sedimentation of fly ash and coolness of warm waste water effluents being received from the TPS. The water pollution tank provided slight sedimentation of fly ash coolness and outer contamination of waste water effluents of TPS. It is indicated by the results of total solides, TDS, TSS, temperature and bacterial count.

At the site IV [From drain (Nallah) area], there was also decrease/increase in various parameters. There was a insignificant ($P < 0.05$) decrease in values of TSS (27.777%) and Mg (33.153%) and significant ($P > 0.05$) decrease in the values of alkalinity (13.684%), Cl (44.827%) and DO (34.267%). However an insignificant ($P > 0.05$) increase in the values of pH (0.05%), free CO_2 (23.809%) and hardness (4.213%) and significant ($P > 0.05$) increase in the value of temperature (65.454%), total solids (684.810%), TDS (150.00%), Ca (16.266%), BOD (51.05%), COD (180.769%), SPC (351.428%) and MPN (2177.777%) were observed in comparison to the control site. At this site the values of all parameters increased/decreased due to the flow rate into the drain route of the waste water effluents; and also due to soil and air contamination of the around areas of the drain route (Tables 5.5–5.12).

The site V indicated further percentage decreases/increases in the values of all parameters. There was further insignificant

($P < 0.05$) decrease in the value of TSS (16.666%) and signifcant ($P > 0.05$) decrease in the value of Cl (41.739%). Besides this, there was insignificant ($P < 0.05$) increase in the values of pH (0.50%), free CO_2 (28.354%), hardness (19.850%) and COD (121.153%) but there was significant ($P > 0.05$) increase in the values of temperature (67.272%), total solids (735.443%), TDS (150.00%), alkalinity (43.157%), Ca (21.394%), Mg (22.170%), dissolved oxygen (37.062%), BOD (120.831%), SPC (397.142%) and MPN (5777.777%) in comparison to site I (Tables 5.5–5.12). At this site the waste water if TPS is being disposed into the field having these of seesum (*Dalbergia sissoo*). The percentage increase in the values of various parameters like solids, dissolved oxygen, BOD, SPC, MPN, free CO_2 and COD (except TSS and chloride) may be due to the increase in number of bacteria, algae, fungi, phytoplankton, protozoa, and coliform bacteria into the drain area and disposal field.

The present study of different physico–chemical and microbiological characteristics of waste water effluents of TPS is important from the point of view that the various parameters like temperature, pH, solids, alkalinity, hardness free CO_2, dissolved oxygen, biochemical (biological) oxygen demand, chemical oxygen demand, calcium, chloride etc. may alter the characteristics of soil where seesum trees are growing. The changes in soil may hamper growth of these trees.

Thus, further study is needed to find out the effect of waste water effluents of thermal power stations on the characteristics of soil and on the growth of various microbes, plants, trees, and agricultural crops.

Summary

In India, coal–based thermal power plants continue to have the maximum share of electric power generated. They produce 35–40 million tonnes of fly ash every year and with increasing time more plants are coming up and this amount is bound to go up increasing the production of this wasteful menace. Therefore the present study was undertaken to determine physico–chemical parameters like, temperature, pH, total solids, total dissolved solids, total suspended solids, alkalinity, carbon dioxide, hardness, calcium, magnesium, dissolved oxygen, biochemical oxygen demand, chemical oxygen demand and microbiological parameters like, standard plate count and most probable number in order to find out the characteristics of waste water effluents of thermal power station, situated in BHEL Ranipur, Haridwar.

Five sampling site viz. Site I/control site (500m before TPS), Site II (100m away from TPS), Site IV (500m for from ash tank) and Site V (1 km far from the ash tank) were selected. The samples were collected every month (July 1995 to November 1995) from each site. The composite samples were used for the determination of various physico–chemical and microbiological parameters following standard techniques.

During present study, the values of physico–chemical and microbiological parameters at site I showed normal characteristics of water. This site showed temperature of 11.0°C, pH of 8.0, total solids of 1.58 ppm, TDS of 0.56 ppm, TDS of 0.72 ppm, alkalinity of 380.00 ppm, free CO_2 of 18.48 ppm, hardness of 221.66 ppm, calcium of 59.016 ppm, magnesium of 17.231 ppm, chloride of 32.94 ppm, dissolved oxygen of 11.594 ppm, BOD of 4.378 ppm, COD of 4.16 ppm, SPC of 17.5×10^4 per ml and MPN of 0.36 per 100 ml.

At the site II, there was an insignificant ($P < 0.05$) decrease in alkalinity, free CO_2, hardness and magnesium and significant ($P > 0.05$) decrease in the chloride, dissolved oxygen, BOD and SPC. But insignificant ($P < 0.05$) increase in the values of P it and calcium and significant ($P > 0.05$) increase in the values of temperature, total solids, TDS, TSS, COD and MPN. The values of parameters decreased/increase in comparison to control site. The changes in value of DO may be due to the closed drain flow and higher concentration of solids (fly ash). Lower values of SPC and MPN may be due to higher temperature while higher value of COD may be due to the different chemicals used during processing.

At site III, the waste–water effluents showed decrease/increase in various parameters. The alkalinity decreased insignificantly ($P < 0.05$) but chloride and dissolved oxygen decreased significant ($P > 0.05$). There was insignificantly ($P < 0.05$) increase in pH, free carbon dioxide, hardness, calcium and magnesium while there was significantly ($P > 0.05$) increase in the values of temperature, total solids, TDS, TSS, BOD, COD, SPC and MPN in comparison to control site. The alteration in values of total solids, TDS and TSS may be due to the sedimentation of fly ash and open flow rate of waste water effluents.

The site IV also showed decrease/increase in various parameters. The values of parameters like alkalinity, chloride and dissolved oxygen significantly ($P > 0.05$) and TSS magnesium insignificantly ($P < 0.05$) decreased. However an insignificantly ($P < 0.05$) increase in the values of pH, free CO_2 and hardness but significantly ($P > 0.05$) increase in the values of temperature, TS, TDS, Ca, BOD, COD, SPC and MPN was observed. The values of all parameters changed due to the fast flow into the open drain route.

At the site V, there was further decrease/increase in the values of various parameters. The values of chloride decreased significantly ($P > 0.05$) and value of TSS decreased insignificantly ($P < 0.05$). The values of pH, free CO_2, hardness, and COD increased insignificantly ($P < 0.05$) and value of temperature, total solids, TDS, alkalinity, calcium, magnesium, dissolved oxygen, BOD, SPC and MPN increased significantly ($P > 0.05$). The reduction in the values of TSS may be due to further sedimentation of solids into the route of drain. The increase in values of BOD and MPN may be due to more microbial contamination.

Finally it was concluded that the increase in alkalinity, hardness, free CO_2 Ca and Mg may be due to an increase in various chemical substances and chemical reactions taking place in the waste water effluents. The increase in DO may be due to increase in the photosynthetic reaction of algae and phytoplanktons of waste water effluents in the open drain. The more contamination of microbes (SPC and MPN) results in an increase in BOD. The decrease in TS, TDS and TSS was due to sedimentation of fly ash in the ash tank. The decrease in COD may be due to separation of chemical in the treatment tanks.

It is a preliminary study. As the waste water effluents are being disposed into fields having plantation of seesum trees (*Dalbergia sissoo*), further study is needed to find out the effect of these effluents on the characteristics of soil and on the growth of various microbes, plants, trees and agricultural crops.

Chapter 6
Sewage Waste Effected

Introduction

Water is one of the prime necessities of life. Man's body contains about 70% of water. It is also an essential requirement for the development of industries and agriculture.

India has abundant natural resources including a number of perennial rivers besides reservoirs, tank, estuaries, stagnant water lakes and swamps. The water from these sources is used for drinking, bathing, sanitation and many other practices (Kudesia, 1991–1992).

Today millions of litres of sewage, domestic wastes, industrial and agriculture effluents containing different harmful substances are being added every year to various water resources. Most of the rivers of world suffer from same problem, disturbing their ecological balance. Man's relentless march towards progress so has been mainly dependent on science and technology that his natural environment stands almost transformed. The human quest for material "development", seriously threatens the fragile ecosystem. Most of our present day environment difficulties can be said to originate from man's "ecological misbehaviour" (Desai, 1990).

Pollution of natural waters by sewage and industrial wastes is objectionable and damaging for many reasons. Of primary importance is the possible hazard to public health and safety; of lesser consequence but still very real are damage to esthetic and intangible attributes of Ganga and other streams and destruction of economic value of clean natural waters (Gurnham, 1955).

The world "pollution" originated from a Latin word "Pollutionem" meaning defilement or to make dirty, and from "polluere" to soil. In this modern age, the problems like population growth, sewage disposal, industrial wastes, radioactive wastes etc., have polluted our natural resources. (Kudesia, 1991–1992).

Holy river Ganga, the most sacred river of India is one of the first ten mighty rivers of the world. It travels nearly 2525 km. through U.P., Bihar and West Bengal supporting about 45 million people. The religious, spiritual, mystical, economic and political importance of Ganga is well known. During its flow through 27 major cities situated at its banks, it became heavily polluted due to discharge of domestic sewage, industrial effluents, incompletely burnt human dead bodies (Sharma and Kaur, 1994–1995; Chopra and Patrick, 1991).

Realising the devastation caused to the population of cities and towns of States and the Union Territory located along the Ganga, the Government of India has launched Ganga Action Plan (GAP) in 1986 to restore its water quality. The river receives about 900 million litres of sewage, chemical and industrial effluents, fertilizers, unburnt and half burnt dead bodies discharged into it. Under GAP (1986) to clean the river and get rid of pollution, nearly 35 sewage treatment plants (STP) were set up and the plan has completed its first phase (Vaidyanathan, 1996). Using these Sewage Treatment Plants (STP), domestic sewage is treated to improve its quality before discharging into river Ganga or on land.

Sewage is a combination of (i) liquid wastes conducted away from residences, institutions and business buildings (sanitary or domestic sewage), (ii) the liquid wastes from industrial establishments (Industrial wastes) with (iii) such surface, ground and storm water as may find its way or be admitted into sewers (known as storm sewage) (Ingram, 1989).

Domesitc sewage consists of discharges of spent water from wash basins, bathrooms and washing machines (Soapy and dirty water), kitchens (food materials) and laboratories (urine, faeces and paper). It is a complex mixture of mineral and organic matter in many forms including (a) large and small particles of solid matter floating and in suspension (b) colloidal and pseudocolloidal dispersion and (c) true solution. Sewage also consists of living matter, specially bacteria, viruses and protozoa; it is an excellent medium for development of bacteria, some of which may be pathogenic (Bolton and Klein, 1971).

If the municipal sewage is directly discharged into natural water bodies like rivers, it will create a high pollution and the river water becomes hazardous for consumption, bathing and other activities. Thus, sewage has to be properly treated before its discharge in river or on land. The treatment of sewage is also necessary when it is used for irrigation and other practices. The treatment of sewage involves following basic steps:

1. **Primary Treatment**: The objective is to remove coarse solids and to accomplish removal of "Settleable" solids.
2. **Secondary (Biological) Treatment**: To adsorb and ultimately oxidize organic constituents of waste water, i.e., to reduce the biochemical oxygen demand (BOD).
3. **Advanced Treatment**: To remove additional objectionable substances to further reduce BOD; includes removal of nutrients such as phosphorons and nitrogen.
4. **Final Treatment:** To disinfect and dispose of liquid effluent.
5. **Solids Processing**: To stabilize solids removed from liquid processes, to dewater solids, and ultimately to dispose off solids for land application, as land fill, and by incineration (Pelczar, Chan and Krieg, 1995).

A number of techniques are employed during the treatment of sewage. Raw sewage is subjected to "Screening" which involves different screens to remove large size floating material from sewage. These materials may otherwise create problems for further treatment processes.

Grit chambers or detritus tanks are used to remove grit, sand and other suspended inorganic solids which may vary from 120 $m^3/10^6m^3$ of sewage (Kshirsagar, 1988).

Intermittent sand filters are used in rural areas but not in major cities due to requirement of large area (Harris, 1977).

Contact beds are water tight compartments having concrete floors and underground drain arrangements for collecting the effluents and then draining it outside. Well operated beds can remove upto 80–90% of suspended matter and 60–70% of BOD.

Trickling or percolating filters utilizes a porous surface like artificial beds of broken stones or coke. The pieces of stone becomes coated with a living film of aerobic, stringly oxidative micro–organisms which oxidize the organic matter of sewage, the result being a much less offensive liquid.

In activated sludge process, flocculated biological growth is continuously circulated and in contact with organic wastes in presence of oxygen for treatment of sewage. During this porcess, colloidal and dissolved organic matter of sewage is changed so that, it is settled down (Mistry *et al.* 1975).

Kshirsagar (1988) opinionated that the sludge can be disposed of by lagooning, burning, land filling or by dumping into sea. Digestion of sludge and its disposal is a part of tertiary treatment of sewage water. The committee of American Public Health Association concluded that "From the hygenic stand point, heat dried activated sludge and heat dried digested sludge appear to be safe, for any reasonable use in agriculture or horticulture. Fresh sludge should be regarded as only one degree removed from the night soil (i.e., excreta) and treated as such, being used only on forage crops and then applied and ploughed in promptly thereafter".

In spite of such expensive and energy consuming techniques, some simple methods like oxidation ponds (stabilization ponds) can also be used for sewage treatment. Oxidation pond or lagoon is an artificial ponds in which , sewage can be retained for sufficient time to satisfy the BOD and thereby, making the sewage non putrescent. According to the type of biological activity occurring in these ponds the lagoons may be of following types :

1. Facultative Ponds of Lagoons

These are most common type of lagoons, sufficiently deep (>2 m) to provide separation into three horizontal stratum. Both anaerobic and aerobic reactions occurs in these ponds. The waste organic matter is suspension are broken down by bacteria releasing nitrogen and phosphorous which are utilized by algae. Such ponds are best suited for small towns and produce an effluent with BOD 30 mg/1 during warm weather operations.

2. Tertiary or Maturation Ponds

These serves for third stage treatment of effluent from activated sludge or trickling filters. Depth is limited to 0.5–1.0 m for mixing and sunlight penetration. Stabilization by retention and surface aeration reduces suspended solids, BOD, faecal micro–organisms and ammonia.

3. Aerated Lagoons

These are completely mixed aerated ponds having basins 3–3.7 m deep. The natural aeration is supported by artificial high speed aerators.

4. Anaerobic Lagoons

These are deep ponds (4.5 m depth) and this treatment is applicable when sewage has high organic strength, high temperature, free from toxic material and sufficient nutrients are available. Various products like CO_2, CH_4, H_2S, and organic acids are produced by micro–organisms. The normal operating standards are to achieve a BOD removal efficiency of 75% on a loading of 320 g BOD/m^3/day, a minimum detention time of 4 days and minimum temperature of 25°C (Hammer, 1986).

Examination of Sewage

Sewage is actually water with a small amount of impurities in it. Examination of sewage is required to know the effects of these impurities. Various tests are used as aid in determining the characteristics composition and condition of sewage (Ingram, W J 1989). Different physico–chemical and microbioloigcal parameters like temperature, pH, total solids (TS), total suspended solids (TSS), total dissolved solids (TDS), hardness, calcium, chlorides, oil and grease, dissolved oxygen (DO), biochemical oxygen demand (BOD), standard plate count (SPC), and most probable number (MPN) are significantly used for estimation of sewage quality, efficiency of treatment process and utility of treated sewage for irrigation and other purposes.

Temperature of water effects biochemical reactions and solubility of gases like oxygen in water samples. The pH of sewage is a measure of acid–base equilibrium achieved by different dissolved components (Webber and Stumn, 1963).

Total solids in water involves all inorganic and organic solids either in suspended or dissolved forms. The percentage of total solids gives an idea about concentration and physical state of sewage.

Dissolved oxygen in water is required for respiration of aerobic micro–organics and depends upon temperature of water sample. BOD is amount of oxygen required by micro–organisms in stabilising organic matter. Acids, alkalies, toxic materials, free chlorine and other bacterial poisons have a marked effect on BOD (Bolton and Klien, 1971). Much of developmental history of BOD upto approximately 1943 has been expertly presented by Phelps (Gaudy, A F Jr., 1972).

Chemical oxygen demand (COD) test is used to measure the content of organic matter in natural and waste waters.

Hardness is caused by calcium and magnesium ions present in water. Hardness of natural waters is comparatively lower than that of sewage (cited in Goel and Trivedi, 1984).

Fats and oils are third major components of food stuffs. The term "grease" as commonly used includes fats, oils, waxes and other related constituents found in waste water.

The science of sanitary water bacteriology began in 1880 when Von Fritsch described *Klebsiella pneumonia* and *K. rhinoscleromatis* as organisms characteristics of human faecal contamination. Later, Escherich identified *Bacillus coli* as an indicator of faecal pollution. Both of them considered human faeces as an indicator of faecal pollution and a dangerous source of contamination (Geldreich, 1966). The MPN tests for total coliforms and faecal coliforms are used to estimate water and waste water quality.

Pathogenic organisms present in water and waste water may cause many serious diseases like typhoid, dysentry, cholera etc. According to Mara (1974), the organisms causing these diseases are highly infectious, therefore, are responsible for many deaths each year in areas with poor sanitation, specially in tropics.

A number of earlier attempts have been made to determine quality of sewage and effect of waste water on natural waters. A large number of sewage treatment plants (STP) were set up in our country for treating municipal and industrial wastes. Thus, it is important to evaluate the treatment efficiency of these plants in terms of various physico–chemical and microbiological parameters. For the same reason, present study was done to estimate the efficiency of STP (sewage treatment plant) at Lakarghat (Rishikesh) and to examine the quality of sewage effluent for irrigation and other beneficial uses after treatment.

Results

The average values for different physico–chemical and microbiological parameters as temperature, pH, total solids (TS), total suspended solids (TSS), total dissolved solids (TDS), turbidity, dissolved oxygen (DO), biochemical oxygen demand (BOD), chemical oxygen demand (COD), chlorides (Cl^-), hardness (as $CaCO_3$), calcium, oil and grease, standard plate count (SPC), most probable number (MPN) for total coliforms and MPN for faecal coliforms at site–I (inlet) and site–II (outlet) for five months (November 1996–March 1997) are depicted in Tables 6.1 and 6.2 respectively. The average values of site–I and II as well as percentage increase/decrease are represented in Table 6.3.

Table 6.1 : Physico–chemical and microbiological parameters of site–I (inlet) of sewage treatment plant at Lakarghat, Rishikesh. Each reading is average of three readings.

Parameter	*November 1996*	*December 1996*	*January 1997*	*Fabruary 1997*	*March 1997*
1. Temperature (°C)	18.6	16.3	14.4	17.2	23.5
2. pH	7.35	8.21	8.21	7.57	7.26
3. Turbidity (NTU)	65.00	60.00	72.00	75.00	68.00
4. Total Solids (mg/litre)	777.00	602.00	755.00	598.00	688.00
5. Total Suspended Solids (mg/litre)	392.00	250.00	366.00	280.00	318.00
6. Total Dissolved Solids (mg/litre)	385.00	352.00	389.00	318.00	370.00
7. Dissolved Oxygen (mg/litre)	Nil	Nil	Nil	Nil	Nil
8. Biochemical Oxygen Demand (mg/litre)	118.00	125.00	88.00	86.00	95.00
9. Chemical Oxygen Demand (mg/litre)	320.00	360.00	240.00	241.00	270.00
10. Chlorides (mg/litre)	54.9	59.6	63.5	53.00	54.00
11. Hardness as $CaCO_3$ (mg/litre)	220.00	260.00	251.00	215.00	225.00
12. Calcium (mg/litre)	48.09	53.5	49.00	56.62	51.5
13. Oil and Grease (mg/litre)	11.60	10.85	9.9	10.2	8.8

contd....

Table 6.1 – contd....

Parameter	*November 1996*	*December 1996*	*January 1997*	*Fabruary 1997*	*March 1997*
14. Standard Plate Count (SPC x 10^5/ml)	66.00	105.00	82.00	69.00	61.00
15. MPN for Total Coliforms (x10^4) Calculated/100ml	220.00	170.00	240.00	350.00	280.00
16. MPN for Faecal Coliforms (x10^4) Calculated/100ml	130.00	140.00	220.00	280.00	240.00

Table 6.2 : Physico–chemical and microbiological parameters of site–II (outlet) of sewage treatment plant at Lakarghat, Rishikesh. Each reading is average of three readings.

Parameter	*November 1996*	*December 1996*	*January 1997*	*Fabruary 1997*	*March 1997*
1. Temperature (°C)	18.7	16.3	14.7	17.2	23.7
2. pH	7.32	7.25	7.75	7.44	7.12
3. Turbidity (NTU)	17.00	18.00	21.00	18.00	19.00
4. Total Solids (mg/litre)	468.00	421.00	450.00	400.0	448.00
5. Total Suspended Solids (mg/litre)	117.00	110.00	85.00	110.00	92.00
6. Total Dissolved Solids (mg/litre)	351.00	311.00	365.00	290.00	356.00
7. Dissolved Oxygen (mg/litre)	1.8	2.3	2.6	2.0	1.7
8. Biochemical Oxygen Demand (mg/litre)	47.00	45.00	42.00	30.00	38.00
9. Chemical Oxygen Demand (mg/litre)	157.00	123.00	154.00	110.00	121.00
10. Chlorides (mg/litre)	40.4	38.00	40.00	38.6	42.00
11. Hardness as $CaCO_3$ (mg/litre)	180.00	192.00	185.00	170.00	198.00
12. Calcium (mg/litre)	41.68	42.8	42.00	43.00	45.00
13. Oil and Grease (mg/litre)	6.5	4.2	4.5	3.8	3.0
14. Standard Plate Count (CFU x 10^5/ml)	48.00	52.00	50.00	44.00	54.00
15. MPN x 104/100ml (Total Coliforms)	180.00	110.00	110.00	220.00	180.00
16. MPN x 104/100ml (Faecal Coliforms)	94.00	94.00	79.00	170.00	110.00

Table 6.3 : Average values of Physico–chemical and microbiological parameters of site–I (inlet) and site–II (outlet) from November 1996–March 1997 and percentage increase/ decrease taking site I as control.

Parameter	*Average Inlet Values*	*Average Outlet Values*	*Percentage Increase/Decrease*
1. Temperature (^{o}C)	18.0	18.12	+ 0.66
2. pH	7.72	7.37	– 4.53
3. Turbidity (NTU)	68.00	18.6	– 72.6
4. Total Solids (mg/litre)	684.00	437.4	– 36.0
5. Total Suspended Solids (mg/litre)	321.2	102.8	– 67.9
6. Total Dissolved Solids (mg/litre)	362.8	334.6	– 7.77
7. Dissolved Oxygen (mg/litre)	Nil	2.08	—
8. Biochemical Oxygen Demand (mg/litre)	102.4	40.4	– 60.5
9. Chemical Oxygen Demand (mg/litre)	286.2	133.00	– 53.5
10. Chlorides (mg/litre)	57.00	39.8	– 30.17
11. Hardness as $CaCO_3$ (mg/litre)	234.2	185.00	– 21.00
12. Calcium (mg/litre)	51.74	42.90	– 17.08
13. Oil and Grease (mg/litre)	10.27	4.40	– 57.15
14. Standard Plate Count (SPC x 10^5/ml)	76.6	49.60	– 35.24
15. MPN (x 104/100ml (Total Coliforms)	252.00	160.00	– 36.5
16. MPN (x 104/100ml (Faecal Coliforms)	202.00	109.40	– 45.84

Note: The negative (–) sign shows a decreases in value as percentage and positive sign (+) shown as increase in value expressed as percentage.

I. Physico–Chemical Parameters

1. Temperature

The temperature at sites I and II was almost same and vary between 14.4^{o}C–23.5^{o}C for site I and 14.7^{o}C–23.7^{o}C for site II during

November 1996–March 1997. The minimum temperature was observed 14.4°C for site I and 14.7°C for site II during January 1997. The Maximum temperature was observed 23.5°C for site I and 23.7°C for site II during March 1997. There is a little increase (+0.66%) in temperature at site II (outlet) in comparison to site I (inlet).

2. pH

The hydrogen ion concentration (pH) as observed for site I during November 1996–March 1997 ranges between 7.26–8.21. pH for site II for same duration vary between 7.12–7.75. The average value of pH (Table 6.3) for sites I and II was observed 7.72 and 7.37 respectively. The decrease in pH was 4.53% between site I (raw sewage) and site II (treated sewage).

3. Turbidity

The turbidity values between November 1969–March 1997 for sites I and II ranges from 60–75 NTU and 17–21 NTU respectively. The average values of turbidity during the study period for sites I and II was observed 68 and 18.6 NTU respectively. An effective decreases (72.6%) was observed at site–II (Table–6.3).

4. Total Solids

The values of total solids (mg/l) for site I ranges from 598–777 during November 1996–March 1997. For site II, these values range from 400–468 mg/l. The maximum and minimum values of total solids for site I and site II are observed during November 1996 and February 1997 respectively. Average value of TS for sites I and II are 684 and 437.4 respectively (Table 6.3). There is significant decrease in total solids between site I (raw sewage) and site II (treated sewage) which is 36% reduction.

5. Total Suspended Solids (TSS)

The total suspended solids ranged from 250–392 mg/l for site I and 85–117 mg/l for site II. The average values of total suspended solids during Nov, 96–March, 97 for sites I and II are 321.2 and 102.8 respectively. TSS decreased significantly (67.9%) at site II (outlet) (Table 6.3).

6. Total Dissolved Solids (TDS)

The dissolved solids ranges between 318–389 mg/l for site I and 290–365 mg/l for site II during the study period. The average values of TDS for sites I and II during November 1969–March 1997 are 362.8 and 334.6 respectively (Table 6.3). There is a little decrease (7.77%) of TDS observed between site I and site II (outlet).

7. Dissolved Oxygen (DO)

No dissolved oxygen was observed at site I during five months (November 1996–March 1997). The DO for site II ranges between 1.7–2.6 mg/l (Table 6.2). The average value of dissolved oxygen at site II was observed 2.08 mg/l (Table 6.3).

8. Biochemical Oxygen Demand (BOD)

The BOD values during study period ranges from 86–125 mg/l for site I and 30–47 mg/l for site II. The average values of BOD for 5 months at site I and site II are 102.4 and 40.4 respectively (Table 6.3). The BOD decreased in significant amount (60.5%) from site I (inlet) and site II (outlet).

9. Chemical Oxygen Demand

The COD values ranges between 240–360 mg/l for site I and 110–157 mg/l for site II during November 1996–March 1997. The average COD values for 5 months at sites I and II are 286.2 and 133.0 respectively. The reducation in COD from site I to site II was observed 53.5% (Tables 6.1, 6.2 and 6.3).

10. Chlorides

The chloride concentration varies between 53.0–63.5 mg/l for site I and 38.0–42.0 mg/l for site II during the study period. The maximum chloride concentration was observed at site I (63.5) during January 1996. The average chloride for 5 months was observed 57.0 and 39.8 mg/l for sites I and II respectively. The average decrease in chloride concentration was 30.17% from site I to site II.

11. Hardness (as $CaCO_3$)

The hardness ranges between 215–260 mg/l and 170–198 mg/l for sites I and II respectively. The average hardness value for sites I and II are 234.2 and 185 mg/l respectively. The effective decreases in hardness from site I to site II was observed 21.0% (Tables 6.1, 6.2 and 6.3).

12. Calcium

The site I (inlet) shows maximum calcium concentration which ranges between 48.09–56.62 mg/l during November 1996–March 1997. The values of calcium ranges between 41.68–45.0 mg/l for site II. The average values for sites I and II are 51.74 and 42.9 respectively. The effective decrease in calcium concentration at site II (outlet) was 17.08% (Tables 6.1, 6.2 and 6.3).

13. Oil and Grease

The values ranges between 8.8–1.6 mg/l for site I and 3.0–6.5 mg/l for site II. The average values observed were 10.27 and 4.4 mg/l for sites I and II respectively. Oil and grease are significantly reduces (57.15%) at outlet site II (Tables 6.1, 6.2 and 6.3).

Results of Microbiological Parameters

1. Standard Plate Count (SPC) (Colony forming units, CFU × 10^5/ml)

The values of SPC are higher at site I (61–105) in comparison to site II (44–54). The average SPC values for sites I and II are 76.6 and 49.6 respectively during the study period. There is a decrease in SPC count (35.24%) at site II in comparison to site I (Table 6.1, 6.2 and 6.3).

2. Most Probable Number (MPN) for Total Coliforms

The MPN values ranges between (170 × 10^4 – 350 × 10^4)/100 ml for site I and (110 × 10^4 – 220 × 10^4)/100 ml for site II. The average values for 5 month were calculated as 252 × 10^4/100 ml and 160 × 10^4/100 ml for sites I and II respectively. The effective decrease in total coliform from site I and site II was 36.5% (Tables 6.1, 6.2 and 6.3).

3. Most Probable Number (MPN) for Faecal Coliforms

The values ranges between 130 × 10^4 – 280 × 10^4/100 ml for site I and 79 × 10^4 – 170 × 10^4/100 ml for site II. The average values of MPN for faecal coliforms are 202 × 10^4/100 ml for site I and 109.4 × 10^4/100 ml for site II. The effective decreases in faecal coliform number was observed 45.84% at site II (Tables 6.1, 6.2 and 6.3).

4. Results of Study of Sewage Microflora

(i) **Algae:** A number of algal species were found to exist in the sewage sample of oxidation ponds. Among the prevailing species, important members identified are

Microcystis sp. (*cyanophyceae*), *Oscillatoria* sp. (cyanophyceae), *Volvox* sp. (chlorophyceae), *Pediastrum* sp. (chlorophyceae) and *Spirogyra* sp. (chlorophyceae) were isolated from oxidation ponds during February 1997 to March 1997. The algae are most prevalent in raw sewage (site I) in comparison to treated sewage (site II).

(ii) **Fungi:** When samples of sewage are plated on Martin's Rose Bengal Agar, a number of fungi appeared on the plates and identified. The most common species observed are *Rhizopus* sp., *Penicillium* sp., *Aspergillus niger*, *Aspergillus flavus* and *Fusarium* sp. The samples for site I shows more crowded plate of fungal population in comparison to the plates of site II.

Discussion

It is important to study the effect of any treatment process applied to improve the quality of sewage in a STP to evaluate the efficiency of plant or the process involved. This was done on the basis of physico–chemical and microbiological characteristics of sewage at different stages of treatment. For the same reason, various characteristics of sewage were examined for site I (inlet) and site II (outlet) at sewage treatment plant, Lakharghat (Rishikesh).

The temperature at sites I and II was almost same, only a little rise of temperature (0.66%) was observed at outlet site in comparison to inlet site. The minimum temperature was observed during January 1997 and maxiumu temperature during March 1997 for both sites I and II (Tables 6.1 and 6.2). Hence, it can be concluded that temperature is largely dependent on climate conditions and there is only minor effect of treatment process on temperature.

The pH of sewage shows that it is slightly alkaline at inlet site and pH decreases in the outlet samples., indicating that sewage water tends to be acidic after treatment. A little decrease in pH was observed (4.53%) at site II in comparison to site I. The pH value of effluent after treatment are within the permissible limits for fish culture and irrigation as per ISI standards (ISI : 2296, 1974).

The solids determined as TS, TSS and TDS were observed very high at site I, showing a heavy pollution load. There is significant decrease in solid contents of sewage as observed at site II (outlet) due to sedimentation and biological treatment processes. According to Rao (1953) and Paka *et al.* (1997), solids in water samples varies directly with alkaline nature of sample. Solids are within the permissible limits of ISI standards (ISI : 2296–1974) for waste water reuse for irrigation.

Turbidity in inlet samples is quite high, having an average value of 68 NTU and is reduces to an average of 18.6 NTU in outlet samples due to sedimentation in the ponds. There is significant reduction in turbidity values during treatment (Table 6.3). Trubidity also varies directly with the solids present is sewage samples.

The value of DO at site I was found zero (Nil) during study period (November 1996–March 1997). Similar results were found at sewage treatment plant, Jagjeetpur, Haridwar (Bhatnagar, 1993), sewage of Udaipur and effluent of dye factories, Pali (Rana and Palria, 1987). The DO value at site I was lower than effluent waste water of STP at Singapore (Koe and Tan, 1990), domestic sewage drain at Jwalapur (Chopra and Rehman, 1991), samples of activated sludge plant in province of Reggio Emilia, Northern Italy (Madoni *et al.*, 1993). Hence, sewage samples of site I in present study show presence of heavy organic pollutants. At site II, maximum DO was observed during January 1997. Do at site I was 'Nil', indicating heavy organic pollution and presence of oxygen demanding wastes. Outlet samples shows a little increase of DO, which may be due to surface aeration and algal photosynthesis. Increase in DO at outlet site causes an adverse effect on BOD, COD, SPC and coliform count as found by Ray and David (1966), and Agrawal *et al.* (1976).

BOD values at inlet site were very high which get reduced in outlet samples due to digestion of organic matter. Average reduction in BOD is found to be 60.5% (Table 6.3) indicating an effective treatment of sewage water. But BOD of effluent was still high as prescribed by ISI (ISI : 4764, 1973) for discharge into inland surface waters.

COD removal observed in present study was 53.5% which is due to oxidation, degradation and stabilization of organic and inorganic wastes. Lens *et al.* (1995) studies a COD removal of 75–95% in domestic waste waters.

The parameters – TDS, chlorides, COD and BOD in present study are lower than sewage at Udaipur city (Rana and Palria, 1987). Hardness (as $CaCO_3$, mg/l) is reduced from 234.2 mg/l to 185 mg/l (Table 6.3) in outlet samples show in an effective reduction of 21%. Calcium is also removed by 17.08% in outlet samples as compared to inlet samples.

Table 6.4 : ISI standards for disposal of waste water. All values are in mg/l, otherwise stated. (Cited in Goel and Trivedi, pp. 240)

Characteristics	*Discharge of Sewage into Inland surface waters*	*Industrial Waste Water in*		*For Inland Surface water when used as Raw Water for public water supplies and bathing ghats*
		Inland surface water	*Public Sewers*	
	IS: 4764–1973	*IS: 2490–1974*	*IS: 3306–1974*	*IS: 2296–1974*
BOD (20°C)	20 p	30.0	500.0	3.0
COD	–	250.0	–	–
pH	–	5.5–9.0	5.5–9.0	6.0–9.0
TSS	30	100	600	–
Temperature °C	–	40	45	–
Oil and Grease	–	10	100.0	–
Sulphide (as S)	–	2.0	–	–
Fluoride (as F)	–	2.0	–	–
Insecticides	–	Zero	–	Zero
Chromium (H.V. as Cr)	–	0.1	2.0	0.05
Nickel	–	3.0	2	–
Selenium	–	0.05	–	0.05
Zinc	–	5.0	15.0	–
Chlorides (as Cl)	–	–	600	600
Sulphates, mg/l	–	–	–	1000
MPN of Coliform/ 100 ml (Monthly Average)	–	–	–	Should not exceed 5000

Oil and grease show a reduction in value from 10.27 mg/l to 4.4 mg/l, i.e., reduced by 57.15% in outlet samples. But the values of oil and grease in treated effluent is higher than standards fixed by ISI for disposal and fish culture (ISI : 2296 : 1974).

The SPC for site I was 76.6 × 105 Colony forming units, (CFU)/ml and for site II was 49.6 × 105 CFU/ml, showing removal of 35.24% an average during the study period. This reduction is due to decrease of organic pollutants during the process of treatment.

Table 6.5 : Tolerance limits for inland surface water for fish culture and irrigation (IS : 2296–1974) (c.f. Mathur, 1993).

Characteristics	*Fish Culture*	*Irrigation*
1. pH Value	6.0 – 9.0	5.5 – 9.0
2. Electrical Conductance at 25°C, Max	1000×10^{-6} mhos	300×10^{-6} mhos
3. Free Carbon Dioxide	6.0	–
4. Free Ammonia (mg/l), Max	1.2	–
5. Dissolved Oxygen, Min	40% saturation or 3 mg/l whichever is higher	–
6. Oil and Greas, mg/l, Max	0.1	–
7. Total Dissolved Solids (Inorganic) mg/l, Max	–	2100
8. Chlorides, mg/l, Max	–	600

The treatment capacity of stabilization ponds relies on the degradation and photosynthesis processes found in stabilization ponds. The non–steady state flow characteristic of these reservoirs is the main factor affecting the reservoir's performance, removal efficiencies and effluent quality (Juanico and Shelef, 1994). In the present study, inlet samples show an average MPN of faecal coliforms as 202×10^4/100 ml which was reduced to 109.4×10^4/ 100 ml in outlet samples. Several studies on the die–off of pathogenic indicators in waste stabilization ponds (total and faecal coliforms) indicate that algal activity and solar radiation are main environmental factors for coliform die–off (Funderburg *et al.*, 1978; Weissman and Kott, 1979; Moller and Calkins, 1980; Trousellier *et al.*, 1986; Pearson *et al.*, 1987). The effective faecal coliform removal in present study was only 45.84%. Dor *et al.*, (1987) found reduction of faecal coliform by one order of magnitude. The high removal of coliform by one order of magnitude. The high removal of coliforms occcurred at high pH values, as studied by Funderburg *et al.*, 1978; Pearson *et al.*, 1987; Saqqar and Pescod, 1990 and Liran *et al.*, 1994).

Various algae found in stabilization ponds in present study include *Microcystis* sp., *Volvox* sp., *Spirogyra* sp., *Oscillatorai* and *Pediastrum* sp. These algae help in nutrient removal from primary settled waste water. Lau *et al.* (1994) reported the uptake of nutrients like organic nitrogen and phosphorous by algae *Chlorella vulgaris.* Algae in water samples also adds oxygen by photosynthesis, hence, improving quality of waste water. This suggests that it is feasible to employ an open algal system to treat primary settled waste water. The intensive algal growth often causes an increase in pH which then leads to ammonia stripping (Talbot and de la Noue, 1993). Rana and Palrai (1987) stated role of *Oscillatoria chlorina* for removal of phosphate nitrate and COD in sewage at Udaipur city.

The fungi in aquatic systems were observed by many workers. In present study, fungi isolated from sewage samples include *Aspergillus niger, Aspergillus flavus, Penicillium* sp., *Rhizopus* sp., and *Fusarium* sp. It is possible that fungal spores from air get deposited at oxidation ponds and germinate there due to presence of high organic matter. Singh and Wadhwani (1987) determined species of *Aspergillus, Penicillium* and *Trichoderma* from tube well waters. No apparent correlation could be established between fungal population and physico–chemical parameters.

The potenital advantages of reuse of treated sewage by irrigation are the low cost of water, beneficial use of plant nutrients in waste water, and use of land application as tertiary treatment. The disadvantages are related to large storage volume of sewage and presence of pathogens which may create health risks. The common coliform standard for restricted irrigation is geometric mean of 5000 coliforms/100 ml with a maximum concentration of 10,000/100 ml in a sample. On the basis of present study, it can be concluded that sewage treatment plant at Lakarghat, Rishikesh receives a heavy load of pollutants and the partial treatment of sewage using oxidation ponds was not sufficient.

Proper treatment of sewage will lead to effluents under permissible limits ofISI and international standards, which has a lot of potential for reuse in agricultural practices, fish culture and industrial reuse.

Conclusion

On the basis of present study, it can be concluded that sewage treatment plant at Lakarghat, Rishikesh receives a heavy pollution load and the partial treatment of sewage using oxidation ponds was not sufficient. Henace, some other processes should also be employed for reduction of various physico–chemical and microbiological parameters of sewage. Following suggestions can help in further improvement of waste water quality :

1. Installation of fine screens to remove larger suspended particles.
2. The use of artificial aerators could increase the DO content of sewage.
3. The use of chemical precipitants will further help in removal of solids in waste water.
4. As the bacterial count goes higher than ISI values for disposal of waste waters, chlorination should be practised for disinfection of water from pathogenic and coliform bacterial population.
5. Continuous monitoring of raw and treated sewage for its quality.

Summary

A number of sewage treatment plants were set up under Ganga Action Plan (1986) to render sewage water harmless before its disposal on land, river Ganga or reuse for irrigation. The treatment efficiency of sewage treatment plant at Lakarghat (Rishikesh) was examined in present study in terms of physico–chemical and microbiological parameters like temperature, pH, total solids (TS), total dissolved solids (TDS), total suspended solids (TSS), turbidity, dissolved oxygen (DO), biochemical oxygen demand (BOD), chemical oxygen demand (COD), chlorides, hardness, oil and grease, standard plant count (SPC), MPN for total coliforms and MPN for faecal coliforms. The stabilization ponds used for treatment were also studied for presence of algae and fungi.

The samples were taken from two sites viz. site I which is main inlet asite of raw sewage and site II is the outlet of last oxidation pond in a series of five ponds in which contents of sewage collection after treatment. The samples were collected and analysed every month from November 1996 to March 1997 and composite samples from each site were analysed according to the standard methods.

Observations during present study reveals that raw sewage at site I shows a heavy pollution load. The average values during a study of five months shows that temperature at site I was 18.0°c, pH of 7.72, total solids of 684 mg/l, total dissolved solids 362.8 mg/ l, total suspended solids of 321.2 mg/1, turbidity of 68 NTU, dissolved oxygen 'Nil', BOD of 102.4 mg/l, COD of 286.2 mg/l, chlorides as 57 mg/l, hardness 234.2 mg/l, calcium 51.74 mg/l, oil and grease 10.27 mg/l, SPC 76.6 × 10^5 cfu/ml, MPN for total coliforms as 252 × 10^4/100 ml and MPN for faecal coliforms as 202 × 10^4/100 ml (Table 6.3).

At site II (treated sewage) a significant reduction was observed for various parameters like total suspended solids, turbidity, BOD, COD and oil and grease. But for other parameters like temperature, pH, total solids, total dissolved solids, DO, chlorides, hardness, calcium, SPC and MPN for coliforms, only a little or medium reduction or increase in values was observed. The increase/decrease in values of different parameters was due to partial treatment of sewage in stabilisation in ponds by sedimentation stabilization and microbial activities in sewage.

The average reduction in values of pH, TS, TDS, TSS, trubidity, BOD, COD, chlorides, hardness, calcium, oil and grease, SPC, MPN (total coliforms) and MPN (faecal coliforms) was 4.53%, 36.0%, 7.77%, 67.9%, 72.6%, 60.5%, 53.5%, 30.17%, 21.0%, 17.08%, 57.15%, 35.24%, 36.50%, and 45.84% respectively. There is little increase in DO due to aerial mixing of oxygen and algal photosynthesis.

Various algae like *Microcystis, Pediastrums, Volvox, Spirogyra* and *Oscillatoria* were observed in the oxidation ponds. They play an important role in sewage treatment by nutrient uptake and algal photosynthesis. Among fungi, different members like *Rhizopus* sp., *Penicillium* sp., *Aspergillus niger, Aspergillus flavus* and *Fusarium* sp. were isolated from sewage sample. But no direct correlation could be established between different parameters and fungal population.

Only pH and solids in effluent waste water after treatment agree with permissible limits of ISI standards for disposal.

Hence, it is concluded that sewage treatment plant at Lakarghat, Rishikesh receives a heavy pollution and is not appropriate for complete treatement of sewge. Hence, it is suggested that various other procedures like application of fine screens, artificial aerators, activiated sludge process, chemical precipitation and disinfection by chlorination should be employed for effective treatment of sewage to make it harmless for disposal.

Properly treated sewage water has a lot of potenital for its reuse in agriculutre, fish culture and industrial application, thereby reducing heavy pollution load on large water bodies like rivers. The present study will certainly help in further designing, modernisation and proper operation of sewage treatment plant at Lakarghat (Rishikesh), which receives the domestic sewage and effluents of small scale industries of Rishikesh.

Chapter 7
Pond Ecology

Introduction

Water is the most vital resource of all kinds of life in this planet. A philosopher and scientist considered water to be the origin of all things. In it about 95% is sea water, 4% is frozen as snow in mountain and only 1% is available for human activity. It is confined in ground water, rivers, lakes, ponds, streams, soil profile and biological system. Out of this ground water constitutes 99% of available water.

Ponds are small bodies of water in which the littoral zone is relatively large and the limnetic and profoundal regions are small or absent. Stratification is of minor importance. Ponds may be found in most regions of adequate rainfall. They are continually being formed for example, as a stream shifts position, leaving the former bed isolated as a body of standing water of "Oxbow". Such ponds can be designated as flood plain ponds since they occur on stream flood plain ponds and are subject to flooding during high water because of the accumulation of organic materials of flood plains, small flood plain ponds may be quite productive. In the piedmont region of the south east, where glacial lakes and ponds are lacking, we have found that flood plain ponds are very favourable sites for class study since they contain a rich and varied life and illustrate various principles and the pond ecosystem. Natural ponds are also formed in limestone region when depressions or "Sinks" develop, due to solution of the underlying strata.

Temporary ponds, that is, ponds which are dry for part of the year, are especially interesting and support a unique community organisms in such ponds must be able to survive, in dormant stage during dry periods or be able to move in and out of ponds, as can amphibians and adult aquatic insects. The fairy shrimps (*Eubranchiopoda*) are especially remarkesable crustaceans which are well adopted to temporary ponds. The eggs survive in dry soil for many months, whereas development and reproduction occur in short time late winter and spring while water is present. Fairy shirmp even occur in temporary pools in the granite on top of stone mountain, Georgia where they are completely isolated vertically as well as horizontally from any other bodies of water. As is true of other marginal habitats, the temporary pond is a favourable place for those organisms adopted to it, because interspecific competition is much reduces. Even though a temporary pond contains water for only a few weeks, a delight seasonal succession of organisms may occur, thus enabling a surprisingly large variety of organisms to utilize a very limited amount of physical habitat.

Artificial ponds of course are the result of damming of stream or basin by man, or by animal such as the beaver price to about 1920, most man made pond in the united state were "mill ponds" formed by damming sizable streams for the purpose of providing power for small mills.

At the present time a very large number of "form ponds" are being constructed which differ from "mill ponds" in that stream water is largely detoured around the pond, in order to prevent loss of nutrients and erosion, or the pond is constructed in basin without a permanent stream. Such ponds have relatively little water flowing through them, and are often artificially fertilized and managed so as discourage rooted vegetation. Thus producers are all phytoplankton and, in a sense, food chain relations resemble those of lakes. However such ponds are rarely deep enough for stratification. A form pond can be much more efficient than an unmanaged pond or mill pond from the stand point of production of a large biomass of fish, but it is less interesting to the naturalist because the variety of organisms is reduced in favour of a large numbers of the desired sps.

Bearer ponds are a characteristic feature of wildness areas of the north and of mountain of west. A bearer pond generally has a short ecological life history, as such ponds are abandoned when the supply of food trees in the vicinity becomes reduced.

Thus, under primeval conditions, the beaver was a very important factor in opening up the forests and in maintain both terrestrial and aquatic seral stage. In the south eastern states and Texas, beaver do not always construct ponds but often live in holes in the band of streams and thus become essentially stream animals.

Let us consider ponds as whole as an Ecosystem. The inseparability of living organisms and the non–living environment is at once apparent with the first sample collected. Not only is the pond a place where animals and plants live, but the animal and plants make the pond what it is. Thus a bottle ful of pond water or a scoop full of inorganic and organic compounds. Some of the larger animals and plants can be separated from the sample for study of counting but it would be difficult to completely separate the myriad of small living things from the non–living matrix without changing the character of the fluid. True one could autoclave the sample of water or bottom mud so that only non–living material remained, but this residue would then no longer be pond water or pond soil but would have entirely different appearances and characteristics.

Temperature influences the number and types of saprophytes, reaeration, O_2 stability and biochemical reaction taking place in a stabilization pond (Davis 1932). A through look into the litreature so for available indicated the paucity of information on the behaviour of individual sps of bacterial and their biochemical role in stabilization pond. *Pseudomonas sps* along with *flavobacterrium* and *achromobacter* constitute about 90–95% of the total bacteria population of the stabilization ponds (Ganapati *et al.*, 1968).

To estimate the total bioactivity of a lake, reservoir of pond. It is necessary to determine the magnitude of primary production and efficiency of energy utilization of different tropic levels.

For these purposes, the estimation of O_2 content in an euphoric zone of any aquatic habitat and indirect estimate of the algal matter seem to offer rapid procedure for determining, through approximately the rate of photosynthesis and energy flow at different trophic levels at various seasons in a year.

Thus for scientific utilization of water resources, the study of phytoplankton and physio–chemical characteristics of water are of primary interest.

The ecological study of Satikund pond has been covering three seasons (winter, summer and monsoon) during 1992–93. The

Satikund pond is situated at Kankhal in the district Haridwar of U.P.

According to old story, in this place the Sati, daughter of Daksha Prajapati and wife of Lord Shiva, burn herself in the Yajna Fires, because she could not tolerate the insults done to her Lord great sanctity is attached to it by the devotees of the Goddess.

Satikund is more or loss square in shape with cemented bathing Ghat that in its western side. It has a 300 × 150 square metre pond with a depth of around 1.75 meters.

It is completely filled with water during the whole year but the maximum water during monsoon and minimum in winter. In its eastern side a large ground is situated. A garden with fullfil of many trees are situated on its North–West. The effluent of some houses also mix in this pond.

It contains a number of phytoplankton including, Diatoms, blue green algae, green algae, etc. and many types of zooplankton. This water is also used in the drinking purposes for cattles and irrigation of land and garden.

Results

The observations of water of the Satikund were made in winter, summer and monsoon seasons during 23 January 1990 – 5 October 1991 are presented in the Tables 7.1 to 7.9.

The water samples of the Satikund were taken once in a month for the following parameters:

Physical Parameters

1. Temperature
2. Transparency
3. Turbidity
4. Total solids (TS)
5. Suspended solids (SS)

Chemical Parameters

6. pH
7. Dissolved oxygen
8. Free CO_2

9. Total alkalinity
10. Hardness
11. Biochemical oxygen demand (BOD)
12. Sodium (Na)
13. Potassium (K)
14. Calcium (Ca)
15. Magnesium (Mg)
16. Chloride (Cl)

Microbiological Parameters

17. Most probable number (MPN)
18. Standard plate count (SPC)
19. Phytoplankton population
20. Zooplankton population

The weather of Haridwar can conveniently be divided into three seasons viz. monsoon from July to October including well demarcated post–monsoon period of September to October. Cold weather period or winter from November to February and hot weather or summer from March to June.

Physical Parameters

1. Temperature

The water temperature of the Satikund was observed to be maximum in the monsoon season (30.4°C ± 0.35) and minimum in the winter season (18.1°C ± 0.60). The average of water temperature recorded at the Satikund was (23.7°C ± 3.4).

2. Transparency

The maximum value of transparency at the Satikund was recorded 63.0 cm ± 4.2 in winter and the minimum value was recorded 20.0 cm ± 1.1 in monsoon season. It showed a decrease after winter (February) due to increase in turbidity. The annual average was 43.0 cm ± 21.6.

3. Turbidity

The turbidity of water was calculated by Nephelometer (Turbidimeter). The lowest value of turbidity at Satikund was recorded 0.5 NTU ± 0.1 during winter season and highest value was observed in summer season as 2.7 NTU ± 0.9. The average value of turbidity was 1.5 NTU ± 1.1.

4. Total Solids (TS)

The lowest value of total solids was observed 202.98 mg/l ± 11.4 in winter season. It showed an increase after winter due to increase in the turbidity. The highest value was observed 390 mg/l ± 18.0 in monsoon. The annual average value was recorded 280.09 mg/l ± 97.2.

4. Suspended Solids

The suspended solids of the Satikund was observed to be lowest in summer season 40.0mg/l ± 5.3 and the highest value was observed 132.0 mg/l ± 12.0 in monsoon season. The average of total suspended solids at this spot was 98.3 mg/l ± 50.8.

Chemical Parameters

6. pH

There was a little variation in pH. During the period of study the lowest value of pH was 8.14 ± 0.05 in the summer season and the highest value was 8.7 ± 0.11 in winter season. The annual average of pH was recorded 8.33 ± 0.30.

7. Dissolved Oxygen

The value of dissolved oxygen of the Satikund was observed to be lowest 8.0 mg/l ± 0.36 in the monsoon period and higher 12.8 mg/l ± 0.38 in winter period. The average of dissolved oxygen for this spot was 10.2 mg/l ± 2.3.

8. Free Carbon Dioxide

The lowest value of free carbon dioxide at this spot was recorded 4.6 mg/l ± 0.78 in the winter season and highest 6.4 mg/l ± 1.1 in monsoon season. The average value of free carbon dioxide was recorded 5.3 mg/l ± 3.9.

9. Total Alkalinity

The total alkalinity at this spot was recorded to be lowest 55.0 mg/l ± 5.7 in the summer season and highest 75.0 mg/l ± 6.0 in monsoon season. The average value of total alkalinity at this sampling station was 67.48 mg/l ± 10.8.

10. Hardness

The total hardness at Satikund was recorded to be lowest 56.0 mg/l ± 7.6 in summer season and highest reading was observed in winter season that is 70.66 mg/l ± 4.6. The average value was recorded 63.88 mg/l ± 7.3.

11. Biochemical Oxygen Demand

The BOD of the Satikund pond was recorded to be lowest 3.6 mg/l ± 0.4 in the winter season, which started increase and the highest reading was observed in monsoon season 4.8 mg/l ± 3.2. The average value of BOD was 4.1 mg/l ± 0.61.

12. Sodium

The lowest value of sodium was observed 1.0 mg/l ± 0.2 in monsoon seasons and highest value of sodium concentration was recorded 3.0 mg/l ± 0.23. The average value of sodium concentration was 2.02 mg/l ± 1.0.

13. Calcium

The calcium concentration of the Satikund was recorded to be lowest 14.07 mg/l ± 0.97 in the winter season. Which started increase and the highest reading was observed in monsoon season that is 20.4 mg/l ± 1.9. The average value of calcium concentration was 17.1 mg/l ± 3.1.

14. Magnesium (Mg)

The minimum Mg concentration recorded was 3.30 mg/l ± 1.1 during monsoon season, which maximum was 3.91 mg/l ± 1.1 in winter season. The average value of Mg concentration was 3.56 mg/l ± 3.3.

15. Chloride

The lowest vlaue of chloride concentration at this spot was recorded 5.6 mg/l ± 0.4 in the summer season and highest 10.85 mg/l ± 0.7 in winter season. The average value of chloride concentration was recorded 9.03 mg/l ± 2.9.

Microbiological Parameters

16. Most probable number (MPN)

The lowest value of MPN at Satikund was recorded 166.6/100 ml ± 0.7 in winter season and highest value of MPN was recorded 300/100 ml ± 0.6 in monsoon season. The average value of MPN stood at 250.5/100 ml ± 1.3.

17. Standard plate count (SPC)

The lowest value of SPC was recorded 9.203×10^3/ml ± 7.8 in summer season. The highest value of SPC was recorded 20.240×10^3/ml ± 9.2. The average value of SPC was recorded 15.867×10^5/ml ± 5.8.

Table 7.1: Physical parameters of the water of Satikund pond during different seasons (1990–91).

S.No. Parameters	Winter	Summer	Monsoon	Average
1. Temperature (°C)	18.1 ± 0.60	23.0 ± 0.34	30.4 ± 0.35	23.7 ± 3.4
2. Transparency (cm)	63.0 ± 4.2	46.5 ± 1.7	20.0 ± 1.1	43 ± 21.6
3. Turbidity (NTU)	0.5 ± 0.1	2.7 ± 0.9	1.3 ± 0.0	1.5 ± 1.1
4. Total Solids (mg/l)	202 ± 11.4	250 ± 14.9	390 ± 18.0	280.9 ± 97.2
5. Suspended Solids (mg/l)	123 ± 9.8	40 ± 5.3	132 ± 12.0	98.3 ± 50.8

± = Standard Deviation

Table 7.2 : Chemical parameters of the water of Satikund pond during different seasons (1990–91).

S.No. Parameters	Winter	Summer	Monsoon	Average
1. pH	8.1 ± 0.11	8.14 ± 0.05	8.0 ± 0.00	8.33 ± 0.30
2. Dissolved Oxygen (mg/l)	12.8 ± 0.38	9.8 ± 0.31	8.0 ± 0.36	10.2 ± 2.3
3. Free CO_2 (mg/l)	4.6 ± 0.78	5.1 ± 1.05	6.4 ± 1.0	5.3 ± 3.98
4. Total Alkalinity (mg/l)	72.45 ± 4.61	55 ± 5.7	75 ± 6.2	57.48 + 10.8
5. Hardness (mg/l)	70.66 ± 4.60	56 ± 7.6	65 ± 5.8	63.88 ± 7.39
6. BOD (mg/l)	3.6 ± 0.40	4.0 ± 0.23	4.8 ± 8.2	4.1 ± 0.61
7. Sodium (mg/l)	2.06 ± 0.208	3.0 ± 0.23	1.0 ± 0.20	2.02 ± 1.0
8. Potassium (mg/l)	1.96 ± 0.05	2.85 ± 0.45	5.6 ± 0.32	3.47 ± 1.89
9. Calcium (mg/l)	14.07 ± 0.97	16.8 ± 2.0	20.44 ± 1.9	17.10 ± 3.19
10.Magnesium (mg/l)	3.19 ± 1.1	3.48 ± 1.2	3.30 ± 1.1	3.56 ± 3.3
11.Chloride (mg/l)	10.85 ± 0.73	5.6 ± 0.49	10.65 ± 0.67	9.03 ± 2.97

± = Standard Deviation

Table 7.3: Microbiological parameters of the water of Satikund pond during different seasons (1990–91).

S.No. Parameters	Winter	Summer	Monsoon	Average
1. (MPN/100ml)	166.6 ± 0.73	285 ± 0.0	300 ± 0.65	250.5 ± 1.3
2. (SPC/ml 10^3)	18.159 ± 11.5	9.203 ± 7.8	2.240 ± 9.2	15.867 ± 5.2
3. Phytoplankton/ml	1207 ± 25.3	1640 ± 22.2	1575 ± 18.8	1474 ± 23.5
4. Zooplankton/ml	182 ± 37.6	206 ± 13.4	240 ± 37.6	209 ± 29.1

± = Standard Deviation

Table 7.4: Quantitative analysis of the plankton of Satikund pond during different seasons (1990–91).

Parameters	*Winter*	*Summer*	*Monsoon*	*Average*
Total Plankton/ml	1389 ± 31.4	1846 ± 17.8	1815 ± 28.2	1883 ± 31.3
Total Zooplankton/ml	182 ± 37.6	206 ± 13.5	240 ± 37.6	209.3 ± 29.1
Total Phytoplankton/ml	1207 ± 25.3	1640 ± 22.2	1575 ± 18.8	1474 ± 33.5
(a) Diatoms/ml	557 ± 28.2	848 ± 38.0	810 ± 22.4	738 ± 29.6
(b) Green algae/ml	462 ± 31.2	560 ± 34.4	495 ± 48.2	505 ± 50.4
(c) Blue–Green algae/ml	188 ± 18.2	232 ± 22.4	270 ± 16.2	230 ± 25.2
Percentage (%)				
(a) Diatoms	40.1	45.9	44.6	43.8
(b) Green algae	33.2	30.3	27.2	30.1
(c) Blue–Green algae	13.5	12.6	14.8	13.6
(d) Zooplankton	13.2	11.2	13.4	12.5

± = Standard Deviation

Table 7.5: State of chlorophyceae (Green algae) present in Satikund pond during different season (1990–91).

S.No.	*Genera*	*Winter*	*Summer*	*Monsoon*
1.	Clostrium	+	+	+
2.	Clamydomonas	+	+	+
3.	Dedogonium	+	+	–
4.	Ankist odesmus	–	+	–
5.	Spirogyra	–	+	+
6.	Ulothrix	–	+	+
7.	Zygnema	+	+	–
8.	Cosmarium	+	–	–
9.	Cladophora	+	–	–
	Total Genera Present	6	7	4

Table 7.6: State of Bacillariophyce (Diatoms) present in Satikund pond during different seasons (1990–91).

S.No.	*Genera*	*Winter*	*Summer*	*Monsoon*
1.	Gomphonema	+	+	+
2.	Synedra	+	+	+
3.	Diatoma	+	+	+
4.	Melosira	+	–	+
5.	Nitzschia	+	–	+
6.	Navicula	+	–	–
7.	Cyclotella	+	–	–
8.	Amphora	–	+	–
9.	Cymbella	–	+	–
10.	Pinnularia	–	+	+
	Total Genera Present	7	6	6

Table 7.7: State of Cyanophyceae (Blue–Green algae) present in Satikund pond during different seasons (1990–91).

S.No.	*Genera*	*Winter*	*Summer*	*Monsoon*
1.	Oscillatoria	+	+	+
2.	Rivularia	+	–	+
3.	Polycystis	+	+	+
4.	Nostoc	+	+	–
5.	Anabaena	+	–	–
6.	Spirulena	+	–	–
7.	Nodularia	–	+	–
	Total Genera Present	6	4	3

Table 7.8: State of Zooplankton present in Satikund pond during different seasons (1990–91).

S.No.	*Genera*	*Winter*	*Summer*	*Monsoon*
1.	Euqlena	+	+	+
2.	Amoeba	+	–	–
3.	Notholca	+	+	–
	Total Genera Present	3	2	1

Table 7.9: Extent of co-efficient of similarity index among phytoplankton between different seasons of the Satikund pond during (1990–91).

		Winter			*Summer*			*Monsoon*		
		1	*2*	*3*	*1*	*2*	*3*	*1*	*2*	*3*
Winter	1				0.46			0.4		
	2					0.46			0.76	
	3						0.6			0.88
Summer	1	0.46						0.72		
	2		0.46						0.66	
	3			0.6						1.0
Monsoon	1	0.4			0.72					
	2		0.76			0.66				
	3			0.88			1.0			

1 = Chlorophyceae 2 = Bacillariophyceae 3 = Cyanophyceae

Plankton

Quantitative analysis of plankton at this spot revealed that the total plankton concentration was highest 1846/ml in the summer season and lowest was recorded in winter season that is 1389/ml. The phytoplankton (1640/ml in summer and 1389/ml in winter) and zooplankton (206/ml in summer and 182/ml in winter) also showed the same type of fluctuation. The annual average of total plankton 1683/ml contributed by both phytoplankton and zooplankton (Table 7.4). The annual percentage of plankton among Bacillariophyceae, Chlorophyceae, Cyanophyceae and zooplankton were found 43.8%, 30.1%, 13.6% and 12.4% respectively.

The phytoplankton concentration at this point dominated by the diatoms (total average 738/ml) followed by green algae (505/ml) and blue green algae (230/ml). The percentage of diatoms was maximum in rainy minimum (40.1%) in winter : of green algae maximum (33.2%) in winter and minimum (27.2%) in monsoon and of Blue–green algae maximum (14.8%) in monsoon season and minimum (12.6%) in summer.

Among bacillasiopgyceae the annual results was shown the presrnce of 10 Genera — *Gomphonema, synedra, Diatoma, Melosira, Nitzschia, Navicula, Cyclotell, Amphora Cymbella, Pinnularia.*

Among chlorophyceae total 9 genera were recorded as *Clostrium, Clamydomonas, Oedogonium, Ankistrodesmus, Spirogyra, Ulothrix, Zygnema, Cosmarium, Cladophora.*

The Cyanophyceae was represented only 7 genera which were *Oscillatoria, Rivularia, Polycists, Nostoc, Anabaena, Spirulina, Nodularia.*

Discussion

The unity of the organism and its environment is a basic principle of Ecology. The organism cannot exist without the environment any organism, population or species lives at the expense of its environment, without this inter action it ceases to exist (Nikolski 1963). The ecological condition of a water body have a direct bearing upon the aquatic life, in the Satikund at Haridwar.

The temperature is one of the most important factors in an aquatic environment. In fact no otehr single factor has profound influence and direct and indirect effect on the biota of an ecosystem. In the present study the minimum water temperature was recorded (18.0°C ± 0.60) in winter season and maximum in monsoon (30.4°C ± 0.35).

Pant *et al.* (1981) observed the same abrupt fluctuation pattern in temperature in lake Nanital, temperature showed an inverse relationship with dissolved oxygen in all seasons as also reported by Gonzalves and Joshi (1985) and Singhal *et al.* (1986).

Sreenivasan (1972), Jindal and Vashist (1985) and Singhal *et al.* (1986). The dissolved oxygen in water is often attributed to the fact that the oxygen is dissolved more during the period of active phytosynthesis. Maximum dissolved oxygen was recorded (12.8mg/l ± 0.38) in the winter season. But the dissolved oxygen reduced onward due to turbidity, which started the photosynthetic activity of flora, the minimum value of dissolved oxygen (8.0 mg/l ± 0.36) was observed in rainy season. This trend is also stated by the Gonzalves and Joshi (1985).

At high temperature during summer and monsoon seasons due to high rate of decomposition of organic matter, release of carbon dioxide resulted in the decrease of pH, as also reported Husmani and Bharti (1980), Yadav *et al.* (1987) also recorded a positive correlation of temperature with free carbon dioxide. Several workers as Zafar (1964). Monawar (1970), Jindal and Vashist (1985), have also reported that free CO_2 influence the alkalinity and pH of water. No correlation could be established between temperature and transparency.

The higher transparency of water (63.0 cm ± 4.2) was recorded in winter season. During summer the transparency decreased and minimum transparency was observed in rainy season which was (20.0 cm ± 1.11). Similar pattern was also reported by Swarup and Singh (1979) in the lake Suraha. The turbidity of water was maximum in monsoon and minimum in water.

The total solids and suspended solids were found maximum in rainy season and minimum in winter. It shows that the turbidity, total solids and suspended matters were closely interrelated with one another and cause common effect upon the pond and its aquatic life as also stated by transparency Khanna (1993).

Maximum BOD was recorded in monsoon and minimum in winter. Highest value of BOD in monsoon was mainly attributed to the highest biological activity at higher temperature whereas the lower BOD in winter indicated lower biological activity at low temperature in winter. This is also revealed by Agarwal *et al.* (1976).

Calcium concentration was found more as compared to that of magnesium concentration or magnesium hardness. The lowest magnesium concentration was recorded in monsoon and highest in winter. It parallel more or less to calcium concentration in its pattern and periodicity. The continuous increase calcium–magnesium concentration of the pond water towards may be attributed to the steady state of hardening of water due to evaporation of the surface water and addition of Ca–Mg salts from detergents and soaps used. This pattern agree with studies of the other workers (Bagde and Varma, 1985, Daborn 1976).

Alkalinity was only due to the presence of biocarbonates. It showed a decreasing tendency during winter to summer, while it increasing thereafter Pant *et al.* (1981) have made the same observations.

The hardness of surface water varied from 56.0 to 70.6 mg/l, the values fluctuating positively in accordance with total alkalinityof water. If is totally depends on alkalinity Chloride ions found maximum 10.85 mg/l ± 0.73 in winter and the minimum 5.6 mg/l ± 0.49 in summer.

The fluctuations in various physico–chemical characteristics may be attributed to, growth, death and decay of phytoplankton in these ponds. Sodium and Pottasium concentration in this pond may be due to better growth in phytoplankton. These findings are conformity with Welch (1951).

Algae present in Satikund pond because of photosynthetic activity, release oxygen into the water and the oxygen is then available to increase aerobic decomposition of the organic wastes of bacteria.

The plankton are in the heterogenous assemblage of minute organisms which occur in the natural water and float about by the wave action and movement of water (Moss, 1982). The present study showed that the maximum total number of plankton were 1846/ml ± 17.8, total phytoplankton was recorded 1640/ml ± 25.2 and zooplankton were 240/ml ± 37.6 in summer season. This minimum concentration of total plankton was recorded to be 1389/ml ± 31.4, phytoplankton 1207/ml ± 25.3 and zooplankton 182.37/ml ± in winter.

The results indicated that the number of plankton were affected due to temperature, content of total solids transparency and turbidity of water.

The same type of results were also reported in a different ponds by Hutchinson (1967) and Das and Upadhyaya (1979). In lakes of Kashmir and Nanital. The maximum percentage of different groups were noted as Bacillariophyceae (45.9%) in summer, Chlorophyceae (33.2%) in winter, and Cyanophyceae (14.8%) in monsoon and zooplankton (13.2%) in monsoon. In this study the Bacillariophyceae was a dominating group as also reported in a pond by Hutchinson (1967).

The number of plankton were also affected by the turbidity. The increased turbidity reduces the plankton production as also reported by Das and Pathani (1978).

The Dissolved Oxygen shows the positive relationship with the plankton but the free carbon dioxide, BOD shows a negative relationship with the plankton.

However a different correlationship between pH and plankton could not be established.

The total count of bacteria were recorded 15.867×10^3/ml ± 5.8 and MPN were 250.5/100 ml ± 1.3 (Table 7.3). It was highest in monsoon season 20.240×10^3/ml ± 9.2 and 300/100 ml ± 0.65 in total count and total coliforms respectively due to high temperature and high percentage of nutrients which increase the growth and multiplication of bacteria.

The bacterial parameters that is MPN and SPC were found high in monsoon. Sagar and Sharma reported the high value in summer and low value in winter. The high value of total count (SPC) and MPN index were also found directly related to organic polllution load. In addition to this some seasonal variation such as temperature turbidity, DO, Free CO_2 and BOD were also responsible upto certain limit.

High rate of residue, Coliform bacteria and SPC in the Satikund pond water so the water is not fit for drinking but it can be used for bathing and irrigation.

Summary

For this study the water samples were collected from three different points in a pond once in each month during 23 January 1990 to 5 Ocotber 1991. The result to water were made in winter, summer and monsoon seasons.

A summary of the finding is given below—

1. The water temperature of Satikund pond at Haridwar ranged between 18.1°C ± 0.60 to 30.4°C ± 0.35.
2. The transparency of water is affected by the suspended impurities and amout of light absorbed. The highest transparency (63.0 cm ± 4.2) of the pond water was observed in winter and minimum transparency (20.0 cm ± 1.1) in rainy season.

3. The turbidity of pond water was almost very low, it ranged between 0.5 NTU ± 0.1 to 2.7 NTU ± 0.9.
4. Residue may affect the water quality, water in high residue generally are in inferior potability. The total solids was maximum 390 mg/l ± 18.0 in rainy season and it was minimum (202 mg/l ± 11.4) in winter season. Suspended solids was maximum (132.0 mg/l ± 12.0) in rainy season and minimum (40.0 mg/l ± 5.3) in winter season.
5. The pH of the Satikund was slightly alkaline. It ranged between 8.14 ± 0.05 to 8.7 ± 0.11.
6. The Chemical parameters which was maximum in winter season were pH (8.7 ± 0.11), dissolved oxygen (12.8 mg/l ± 0.38), hardness (70.66 mg/l ± 4.60), magnesium (3.91 mg/l ± 1.1) and chloride (10.85 mg/l ± 0.73). And parameter which was in minimum in winter season were free carbon dioxide (4.6 mg/l ± 0.78), BOD (3.6 mg/l ± 0.40), potassium (1.96 mg/l ± 0.05) and calcium (14.07 mg/l ± 0.97).
7. Sodium is the only single chemical parameter which was maximum in summer season (3.0 mg/l ± 0.23). The chemical parameter which was minimum in this season were pH (8.14 ± 0.05), total alkalinity (55.0 mg/l ± 5.7). Hardness (56.0 mg/l ± 7.6) and chloride (5.6 mg/l ± 0.49).
8. In monsoon season the chemical parameters which was maximum were free CO_2 (6.4 mg/l ± 1.1), total alkalinity (75.0 mg/l ± 6.2), BOD (4.8 mg/l ± 3.2), potassium (5.6 mg/l ± 0.32) and calcium (20.44 mg/l ± 1.9) sodium (1.0 mg/l ± 0.20) and magnesium (3.30 mg/l ± 1.1).
9. The total coliform organism (MPN) of the Satikund pond at Haridwar ranged between 166.6/100 ml ± 0.73 to 300/100 ml ± 0.65. The maximum MPN was in rainy season and minimum in winter season.
10. The total count of bacteria (SPC) of the pond water ranged between 9.203 × 10^3/ml ± 7.8 to 20.240 × 10^3/ml ± 9.2. The maximum SPC was in rainy season and minimum in summer season.

11. The index of similarity among different genera of diatoms, green algae and blue green algae showed that it was below 1.0 through out the year.
12. There was a positive relationship between plankton and dissolved oxygen.
13. There was a negative relationship between plankton and BOD.
14. The genera of phytoplankton and zooplankton of the Satikund pond at Haridwar are listed in the test.

    Total 26 genera of phytoplankton were observed. The phytoplankton concentration was dominated by the diatoms (*Gomphonema, Synedra, Diatoma, Melosira, Nitzschia*). Total 3 genera of zooplankton were recorded among with euglena showed its dominance.
15. Water is not fit for drinking due to high value of residue, Coliform bacteria and SPC. But is can be use bathing and irrigation.

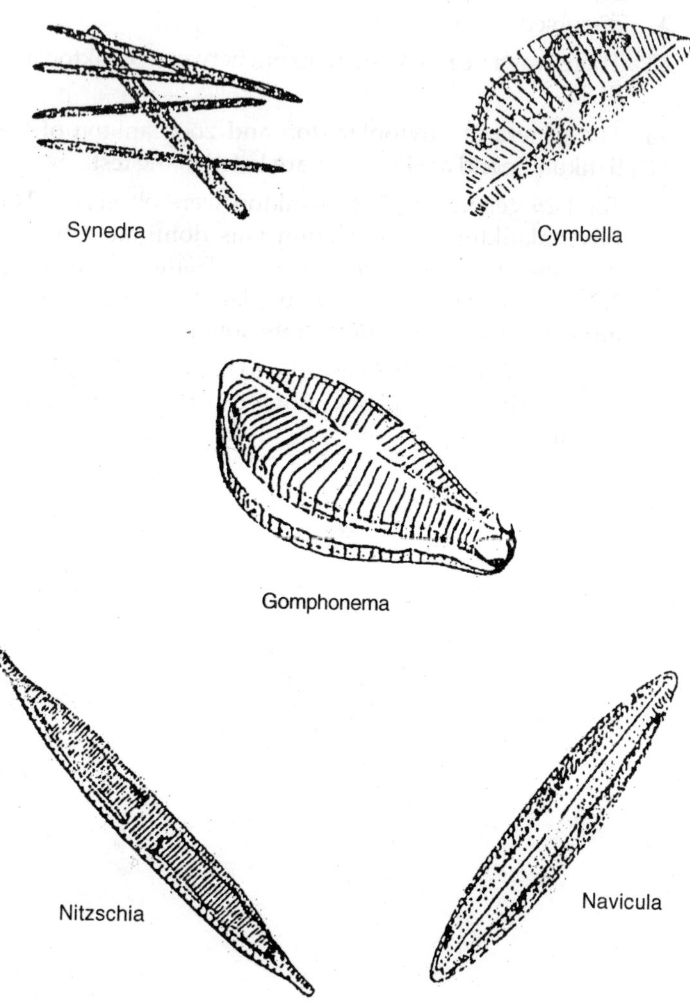

Fig. 7.1: Different Genera of Bacillariophyceae (Diatoms)

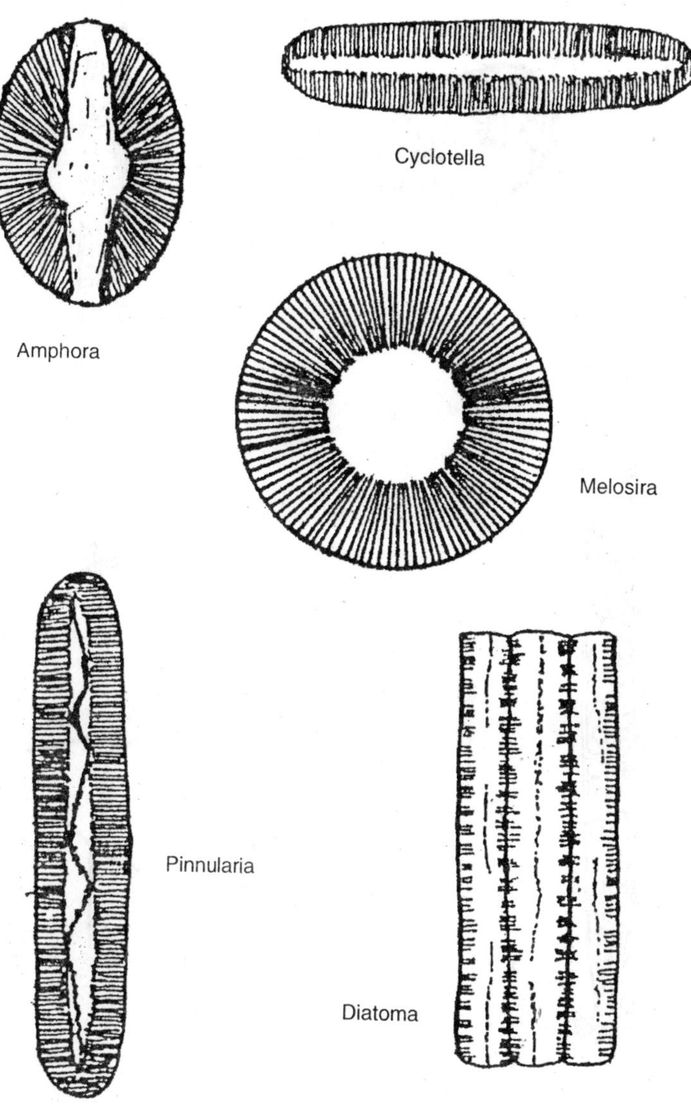

Fig. 7.2: Different Genera of Bacillariophyceae (Diatoms)

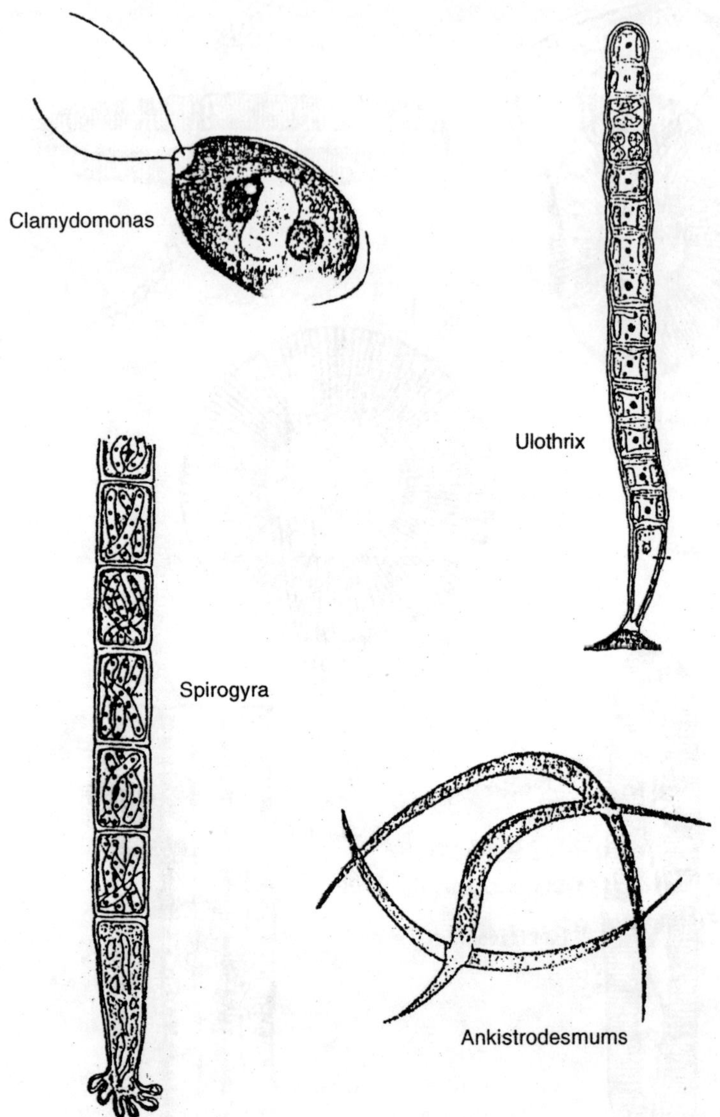

Fig. 7.3: Different Genera of Chlorophyceae (Green algae)

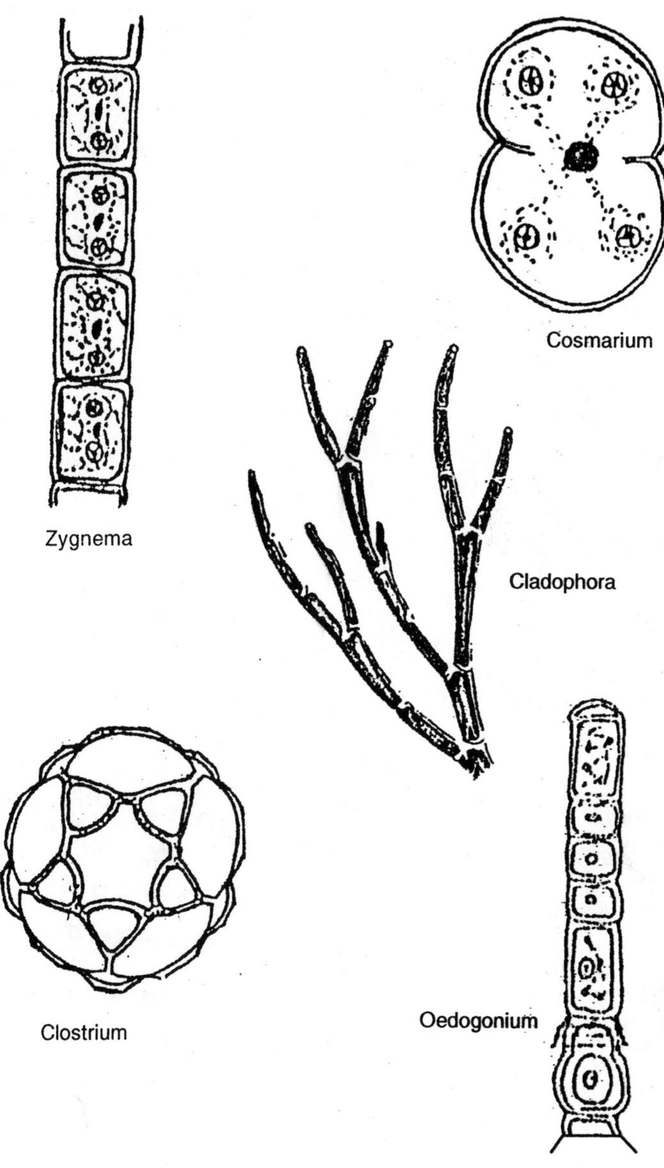

Fig. 7.4: Different Genera of Chlorophyceae (Green algae)

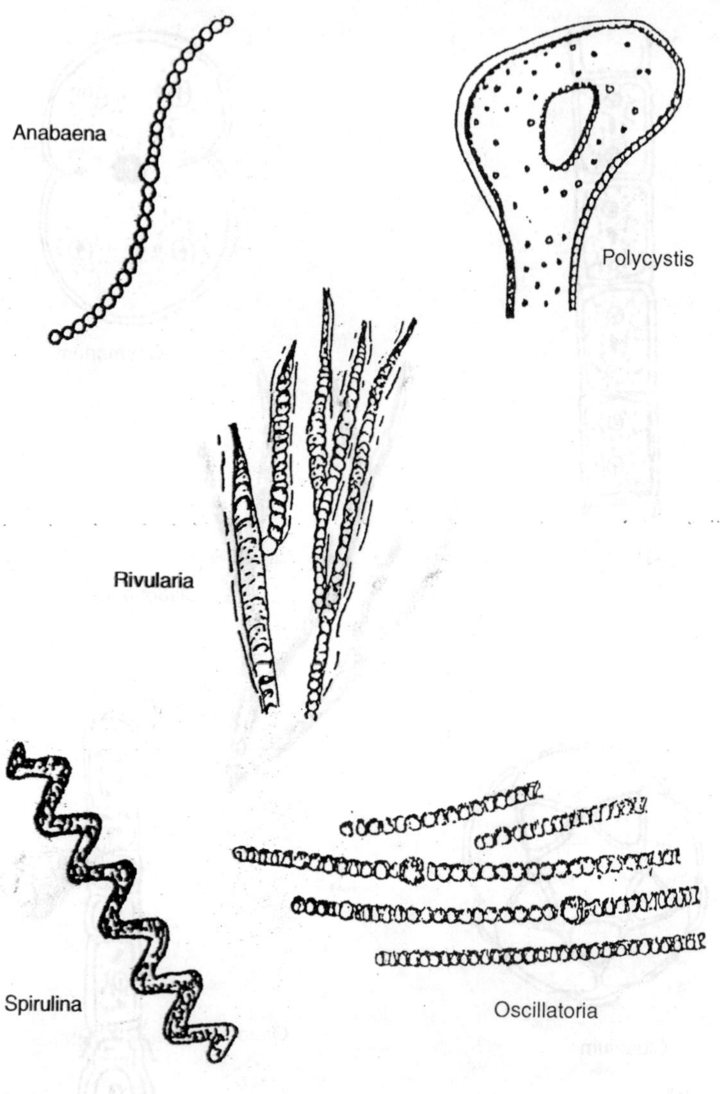

Fig. 7.5: Different Genera of Cyanophyceae (Blue–Green algae)

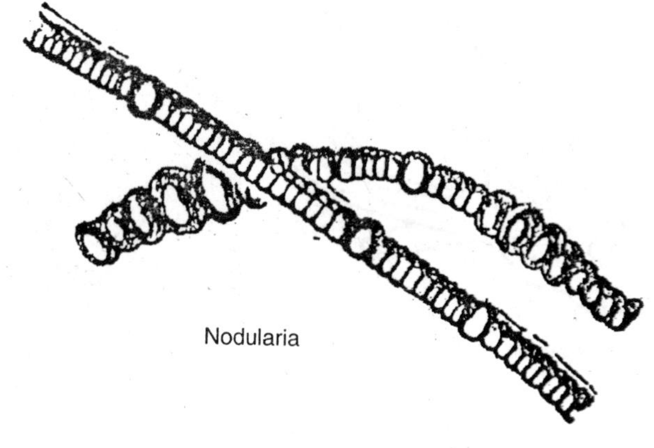

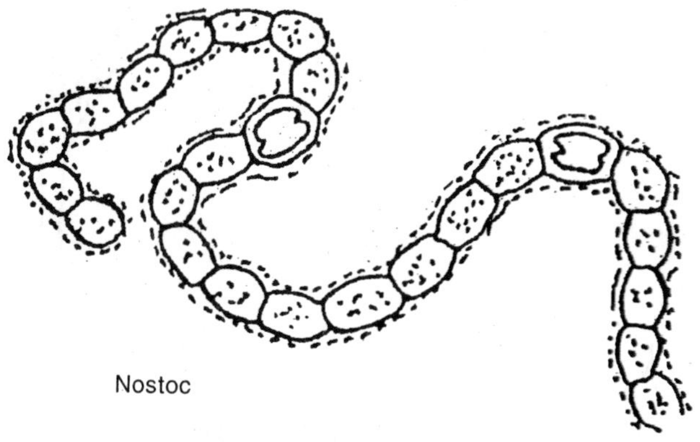

Fig. 7.6: Different Genera of Cyanophyceae (Blue–Green algae)

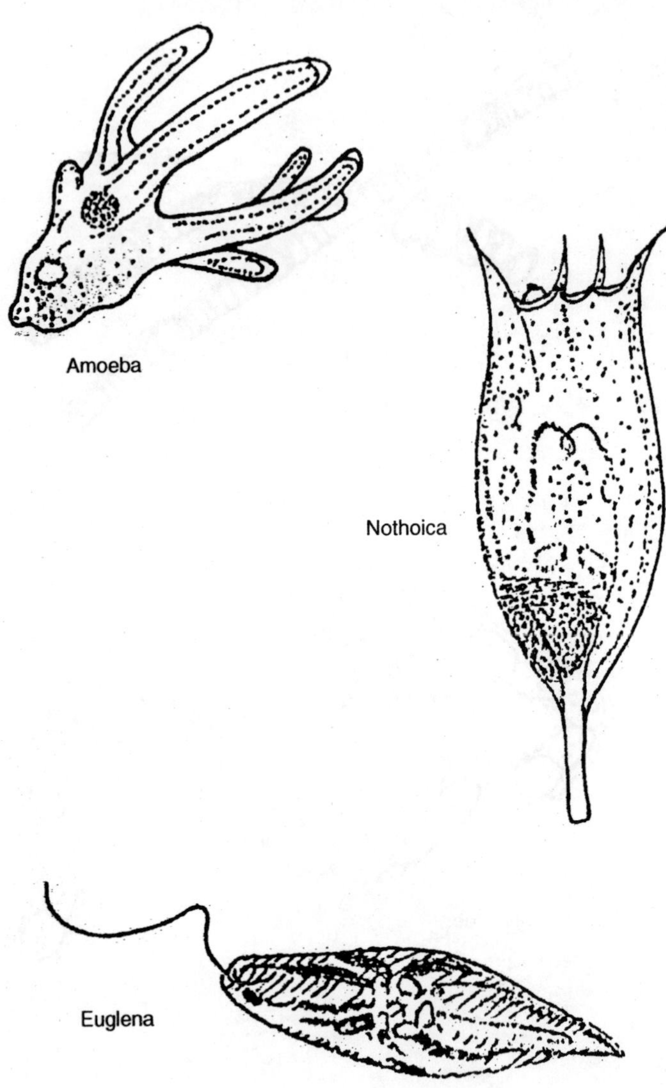

Fig. 7.7: Different Genera of Zooplankton

Bibliography

Abdin, G. (1948). Physical and Chemical investigations relating to algal growth in the river Nile, Cairo Bull. Inst. Egypt., 29: 20–24.

Abdin, G., (1948b). Seasonal distribution of Phytoplankton and sessile algae in the river Nile. Cairo.*Ibid.*, 29:369–382.

Abeliovich, Aharon.(1985): Biological treatment of Chemical Industry Effluents by stabilization ponds". Water Res. Vol.19, No. 12: pp. 1497–1503

Adebisi, A.A. (1981). The Physico–chemical hydrology of a tropical seasonal river upper Ugur river hydrobiologia, 79.

Adholia N. Upkar (1991). Primary productivity of Mansarovar Reservoir, Bhopal Him. J. Env. Zool., 5, 1991: 79–81.

Agarwal, D.K. Gaur and S.D., Tiwari, I.C. Naraganaswami, M.S. and Harwai, S.M. (1976.) Physico–chemical characters of the Ganga at Varanasi India J. Environ. Health, 18 : 201–206.

Alexander, M. (1978). An Introduction to soil microbiology. John Wiley and sons, New York. pp. 21–27.

Anand, V.K., Sharma, S. and Anand, K. (1993). Aquatic and Marshy plants of Jammu province II – systematic enumerations. Advances in plant sciences, Vol VI (1).

Anderson, G.E., Comita, G.W. and Engstrom, Reg. V. (1955). A note on Phytoplankton–zooplankton relationship in two lakes in Washington. Ecology, 36: 757–759.

Anderson, R.R. (1969). Temperature and rooted aquatic plants. chesapeake sci 10, 157–164.

Andreadaksi, A.D. (1993). Physical and chemical properties of activated sludge floc. Wat. Res. Vol. 27 (12): 1707–1714.

Anonymous (1988). "Hindu", September 1–8–1988.

APHA (1985). In: Standard methods for examination of water and waste-water, 16th edn. APHA, AWWA, WPCF, Washington.

Arokiasamy, D.I., Meenakshi, R.M. and Gananarethiam 1981. Effect of distillery effluent on water potential of the rice seedlings. Indian J Exp. Biol., 19: 96–98.

Arora, H.C., Routh, Tapan, Chattopadhya, S.N. and Sharma, V.P. (1974). Survey of sugar mill effluent disposal part II. A comparative study of sugar mill effluent characteristics. Indian J. Environ. Health. 16:233.

Badala, S.P. (1979). Ecological studies on the Khthyo–fauna of some fresh water resource of Garhwal region. D.Phil thesis, Garhwal University.

Badala, S.P. and Singh, H.R. (1981). Hydrobiology of the river Alaknanda of Garhwal Himalaya Indian J. Ecol, 8(2): 269–276.

Bagchi, K.G., 1986. Integrated eco–development project on river Ganga basin project submitted to Department of environment. Government of India. Calcutta University, Calcutta.

Bagde, U.S. and Varma, A.K. (1985). Physico–Chemical characteristics of water of J.N.U. lake at New Delhi. Indian J. Ecol. 12(1): 151–156.

Baker, A. (1961): "Taste and odor studies in water". J.Water pollution country Fed. 33: pp. 1066–1099.

Balakrishnan, P.K. and Joseph, Amini (1990): "Effect of Industrial discharge on the Ecology of phytoplankton production". Wat. Res. vol. 24. No. 6: pp. 787–796

Balmer, P. and Hultman, B. (1988). Control of phosphorous discharges: present situation and trends. Hydrobiologia. 170: 305–319.

Balone, E.K. and Coche, A.G. (1974): Lake Karina. a man made tropical ecosystem in Central Africa. Mangrbiol 24: 767.

Baloni, M.C. and Sarkar, H.L. (1965). Since on the pollution of Yamuna river at Okhla water works intake Delhi, Indian J. Environ. Health 7: 84–82.

Balusu, K.R. & Sharma, U.P. (1966): "Survey of Najafgarh drain down of industrial area". Indian J. Environ Health. vol. 8: pp 103–111

Bansal, Samidha, and Khare, G.K. (1990). River pollution due to paper industry. A case study. Indian J. Env. Protec., 10(4) : 281–283.

Baruah, B.K., Baruah, D. and Das, M. (1996). Study of the effect of paper mill effluent on the water quality of receiving wetland Pollu. Res., 15(4): 389–393.

Bauman, E.R. (1965). In physical scientific and engineering aspect of pollution ch–2 water pollution control and abatment. Iowa state Uni. Pr., pp. 29–39.

Beak, T.W.(1964): "Biological measurement of water pollution". Chem. Eng. Papers–60: pp 33–45.

Bernard, Claude, Regnet, Yves & Colin, Francois. (1987): "Automatic detection of coliform bacteria for Industrial control of Drinking water quality". *Wat. Res.* vol. 21. No. 9: pp. 1089–1099.

Berner, L.M. (1951). Limnology of the lower Missouri River, ecology 32: 1–12.

Bhargava, D.S. (1985). "Variations in the equality in the Ganges". J. (Effluent and water treatment Journal). pp 33–41.

Bhargava, D.S. (1985). Risk assessment for pilgrims. Journal of the Institution of Engineers. 65: 99–100

Bhargava, S.E. and Alam, M. (1979). Seasonal succession of Diatoms in relation to certain Physico–chemical factors at Sambhar salt lake and its reservoir, Geobios, 6: 207–211.

Bhaskaran, T.R., Chakraborty, R.N. and Trivedi, R.C. (1963). Studies on the river pollution 1. Pollution on Purification of Gomati river near Lucknow. Journal of the Institution of Engineers. 45: pp. 39–50.

Bhatnagar, M.K. (1993). A study of physico–chemical properties of industrial effluent and city/sewage waste effected by biogas production. G.K.V.V., Haridwar.

Bhatt, S.D., Bisht, Yashodhara and Negi, Usha (1985). Ecology and phytoplankton in river Kosi of the Western Himalaya (U.P.) Indian J. Ecology 12 (1): 141–146.

Bhowmick, B.N. and Singh, A.K. (1985). Phytoplankton population in relation to physico–chemical factors of river Ganga at Patna. Indian J.Ecology. 12 : 360–364.

Bilgrami, K.S. and Datta Munshi, J.S. (1979). Limnological survey and impact of human activities on the river Ganges (Barauni to Farakka range). A Technical report. Post–Graduate Deptt. of Botany, Bhagalpur University, Bhagalpur.

Birge, W.J. (1978). In: Aquatic toxicology of trace elements of coal and fly ash. Energy and environmental stress in aquatic systems. (edn. Thro, J.H. and Gibbons, J.W.), 219–240. Symposium Augusta, Georgia, 1977. U.S. Department of Energy.

Bist, K.L. and Chopra, A.K., 1992. Studies on seasonal Diurnal rhythm of some physico–chemical qualities of the river pinder of Garhwal Himalaya. Him.J.Env.Zool. 6: pp. 172–175.

Boissya, S.L. and Dutta, S.K. (1996). Effect of paper mill effluent on germination of its seed Oryza sativa (var. masuri) and growth behaviour of its seedlings. J. indl. poll. Cont. 12(2): 123–128.

Bolton, R.L. and Killen L. (1971): Sewage Treatments: Basic principles and Trends, Butter worth and Co. Ltd. pp. 9–37.

Borrego, J. Juan and Morinigo, A. Miguel (1990): "Coliphages as an indicator of fecal pollution in water. Their survival aquatic Environment". Wat. Res. Vol. 24. No. 1: pp. 111–116.

Bouwer, H. (1978). Ground–water Hydrology. McGraw. Hill Inc., Tokyo and Japan.

Bruvold, W.H. and Ongerth, H.J. (1969). Taste quality of mineralized water. Journal of AWWA., 61: pp. 170.

Byrene, C.D., Law, R.J. and Thain, J.E. (1988): "Measurement of the dispersion of Liquid industrial waste discharged in to a wake of a dumping vessel" Water. Res. Vol. 22. No. 12: pp. 1577–1584.

Cairns, J.Jr. (1956). Effect of increased temperatures on aquatic organisms. Industrial wastes 1, 150.

Calman and Dixit, A.D. (1989). Change in water quality of industrial waste disposal; Indian J. of Environ. Health. 31(1): 73–78.

Camp, T.R. and Meserv, R.L. (1974): "Water and its impurities" Second Ed. Dowden, Hutebinson & Ress Pannsylavania.

CBPCWP (1985): "Adsorption of treated or untreated organic load by the soil in term of BOD & COD" Indina Jour. Environ. Health. Vol. 8. No. 1: pp. 131–135.

Chacko, P.I and Srinivasan, R. (1955). Observations on the hydrobiology of major rivers of Madras State. South India, fresh Biol. Stn. *Modrus*, 13 : 1–14.

Chacko, P.I. and Ganapati, S.V. (1949). Some survey on the Adyar River with especial reference to its hydrobiological conditions Indian Geogr. J., 24 (3).

Chakraborty, R.D., Ray, P. and Singh, S.B. (1959). A quantiative study of the plankton and the physico–chemical condition of river Jamuna at Allahabad, 1954–55. Indian Jr. Fish. 6: pp. 186–203.

Chattopadhya, S.N., Routh, T., Sharma, V.P., Arora, H.C. and Gupta, R.K., (1984). A short–term study on the pollutional status of river Ganga in Kanpur region. Indian J. Environ.Hlth.3: 244–257.

Chaudhary, N.K., Pradhan, P.K and Desh, M.C. (1979): "Certain physico–chemical factors and phytoplankton of Kiradudua Damp". Geobios, Vol. 6. : pp. 104–106.

Chaudhary, S. (1986). Integrated eco–development of the river Ganga basin projects (Consolidated progress report). Bidanchandra Krishi Vishwavidyalaya, Kalyani.

Chaudhary, S.K. Jha, A.M. and Shrivastava, D.K. (1987). Effect of paper mill effluent and on seed germination and seedling growth of Zeamays. Environ. and Ecology. 5(2): 285–287.

Chaudhary, S.K., Jha, A.M. and Shrivastava, D.K. (1989). Impact of paper mill effluent on germination seedlings growth and pigment of Hordium–Vulgare; Environ. and Ecology, 7(1): 193–195.

Cherry, D.S. and Gutherie, R.K. (1979). The uptake of chemical element from coal ash and settling basin effluent by primary producers. II. Relation between concentration in ash deposits tissues of grass growing on the ash. Sci. Total Environ., 13, 27–31.

Cherry, D.S., Guthere, R.K., Sherberger, F.F and Larrick, S.R. (1979). The influence of coal ash and thermal discharge upon the distribution and bioaccumulation of aquatic invertebrates. Hydrobiologiai,62, 257–267.

Chona, M.K. (1990): A report of efficiency of primary treatment plant, Chandigarh. Him. J. Env. Zool., Vol 4. : 173–176.

Chopra, A.K. and Patrick, Nirmal , J.C. (1991): Effect of domestic sewage on water quality of river Ganga at Triveni Ghat, Rishikesh, G.K.V.V.

Chopra, A.K. and Patrick, N.J. (1994). Effect of domestic sewage on self–purification of Ganga water of Rishikesh. 1. Physico–chemical Parameters. Ad. Bios. 13(1): pp. 75–82.

Chopra, A.K. and Rehman, A. (1991). Pollution of waste water entering upper Ganga canal at Jwalapur, Haridwar. J. Ecobiol. 4: 145–149.

Chopra, A.K., Madhwal, B.P. and Singh, H.R. (1990). Abiotic variables and primary productivity of river Yamuna at Naugaon, Uttarkashi–Garhwal, Ind. J. Ecol. 17(1): 61–64.

Clapp, C.E., Larson, W.E., Dowdy, R.H., Linden, D.R., Marten, G.C. and Duncomb, D.R. (1983). Utilisation of Municipal Sewage Sludge and Waste Water Effluent on agricultural land in Minnesota. Proc. II, Intern. Symp. on Agriculture and Horticulture.

Cosey, H. and Neton, P.V.R. (1973). The chemical composition and flow of the river Frome and its main tributaries. Freshwat. Biol. 3 : 317–333.

Curds, C.R. (1992): Protozoa in Water Industry. Cambridge University Press, Cambridge, U.K.

Daborn, G.R. (1976). Physical and chemical factors of vernal temporary pond in Wastern Canada, Hydrobiologia S1(1): 33–38.

Daniel, B. and Rajamani, S. (1992). Characterization of indicator bacteria in recreational water supplies. Advances in Plant Sciences, 5(2): 448–456.

Das and Pandey, J. (1978). Some physico–chemical and biological indicators of pollution in Lake, Nainital, Kumaon (U.P.) Indian J. Ecal 5: 7–16.

Das, S.M. and Akhtar, S. (1940). A report on fresh water plankton from the Dal Lake Kashmir. Kashmir Sc., 7(1–2): 133–137.

Das, S.M. and Pathan, S.S. (1978). A study on the effect of lake ecology on productivity of mahaseer (Tor–tor and tor–putitora) in Kumaon lakes, India, Matsya, 4: 25–31.

Das, S.M. and Srivastava, V.K. (1956). Studies on fresh water plankton II Correlation between plankton and hydrological factors Proc. Nat Acad Sci India (B) 26: 243–254.

Das, S.M., Upadhyaya, J.C. (1979). Studies on qualitative and quantitative fluctuations of plankton in two Kumaon lakes, Nainital and Bhimtal (India). Acta. Hdrobiol, 21(1): 9–17.

Datta, N.C., Mandal, N. and Bandyopadhyay, (1987). Seasonal abundance of Rotifers in a perennial fresh water pond in Calcutta. J. Env. Biol., 8(1): 63–71.

David, A. (1956). Studies on the Pollution of the Bhadra river at Bhadrawati fisheries effluents. Pro.Nat. inst. set India, 93(3): 132–160.

David, A. (1963). Report of fisheries, survey of the river Gandak (North Bihar) Surv. Reo. Cent. Ind.fish. inst. Brrak 1–24.

Davis, C.C. (1948): "Studies on the effect of industrial pollution on the lower patapasco river area. The effect of copper pollution on plankton". Chesapeake Biog. Lab. Pub. Bo. 72: pp. 11–12.

Davis, N.C. (1932). The effects of heat and cold upon Aedes (Stegomyia) aegyptil Amer. J. Hyg. 16. 177–191.

Delfino, J.J. and Byrras, D.J. (1975). The influence of Hydroloigical conditions on dissolved and suspended constituents in the Missouri River water. Soil Pollut, 5 (2): pp. 157–168.

Desai, P.V. Godase, S.J. and Halkar, S.G. (1995) Explain. The physico–chemical. Characteristics of Khandepur River, Goa, India Roll Res. 14 (4): 447–454.

Deshmukh, S.B., Phadke, N.S. and Kothanandaraman (1964). Physico–chemical characteristics of Ambazari lake water. Environmental. Health, 6: 186–188.

Deshmukh, S.B., Phadke, N.S. and Kothandareman, V. (1964). Physico–chemical characteristics of Kanhan river water, Nagpur City. Indian J. Environ. Hlth. 6: pp. 181–183.

Dobriyal, A.K. (1985). Ecology of Limnofauna in the small streams and their importance to the village life in the Garhwal Himalaya. Uttar Pradesh J. Zool. 5(2): 139–144.

Dobriyal, A.K. and Singh, H.R. (1981). Diurnal aeration in some aspects of limnology of the river Mandakini from the Garhwal Himalaya. Uttar Pradesh J. Zool 1: 16–18.

Dreeson, D.R., Glandney, E.S., Ownes, J.W., Perkins, B.L., Wienke, C.L. and Wange, L.E. (1977). Comparison of levels of trace elements extracted from fly ash and levels found in effluent waters from a coal fired power plant. Environ. Scie. Technol., 11, 1017–1021.

Duzzin, B., Pavoni, B. & Doonozolo, R. (1988): " Macro–invertebrate communities and sediments as pollution indicator for Heavy metal in the river Adige (Italy) Receiving Industrial effluents". Water. Res. Vol. 22. No. 1: pp. 1353–1353.

Edward, C.A.(1977): "Persistent pesticides in the Environment". C.R.C. Press, cleveland Ohio.

Eloranta, P.V. (1983). Physical and chemical properties of pond waters receiving warm water effluent from a thermal power plant. Water Res., 17, 133–140.

Finch, G.R. and Smith, D.W. (1986). Batch coagulation of a logoon for faecal coliform reductions. Wat. Res. Vol. 20(1): 105–112.

Forbes, S.A., and Richardson, R.E. (1913). Studies on the biology of the upper Illinois River. Bulletin of Illinois State Laboratory of Natural History, 9: pp. 481.

Freeman, Martin and Jaggi, B. (1986). Pollution performance of firm from pulp and paper industry J. Environ. Manage. 10(3): 359–366.

Fugiya, M. (1961). Effect of kraft pulp mill waste water J. water poll. Cont. Fed., (33): 968.

Funderburg, S., Moore, B., Sorber, C., and Sagik, B. (1978). Survival of poliovirus in model waste water holding ponds. Prog. Water. Technol. 10(5/6): 619–629.

Furr, A.K., Kelly, W.C., Bache, C.A., Gutenmann, W.H and Listk, D.J. (1976). Multi–element uptake by vegetables and millet grown in pots on fly ash amended soil. J. Agricult. Food Chemistry, 24, 885–888.

Ganapati, S.V. (1957). Limnological studies of two upland waters in Madras state. Arch Hydrobiol 56: 30–61.

Ganapati, S.V. and Sreenivasan A (1968). Aspects of limnology, primary production and fisheries in the Stanley reservoir Madras state. Hydgobiol 321: 551–569.

Ganpati, S.V. and Alikunhi, K.H. (1950). Factory Effluents from the mettur chemical and industrial corporation Ltd. Metter Dam Madras and their pollutional effects on the fishes of the river Cauvery. Proc. Nut. inst. Sci.India 16 (3) : 189.

Ganpati, S.V. and Chacko, P.I. (1957). An investigation of the river Godavary and the effect of Paper Mills Pollution at Rajan Mundary Proc.I.P.F.C. sec. 11:5.

Garrote, J., Castro, Pablo and Bao, J. Manuel (1995): "Treatment of Tannery effluents by a two step coagulation Flocculation process". Wat. Res. Vol. 29. No. 11: pp. 2600–2608.

Gaudy, A.F. Jr. (1972): Water pollution Microbiology by Ralph Mitchell, John Wiley and Sons. pp. 305–331.

Geldreich, E.E. (1966). Sanitary significance of faecal coliforms in the environment, Water pollution control research series. Publ. Wp–20–3, FWPCA, USI, Cincinnati, Ohio.

Genjatulin, Valeevich Karl (1990): "Controlling chemical and biological water pollution by Quantitative bioassaying". Wat. Res. Vol. 24. No. 5: pp. 539–541.

Gersberg, R.M., Elkins, B.V., Lyon, S.R. and Goldman, C.R. (1986). Role of aquatic plants in waste water treatment by artificial wetlands. *Wat. Res.* Vol 20(3): 363-368.

Ghatak, D.B. and Konnar, S.K. (1992). Field study on the effluent of various industrial effluent on Damodar river Ecosystem West Bengal. Environ. and Ecology 10(4): 767–773.

Ghatak, D.B. and Konnar, S.K. (1992). Impact of various industrial effluents on Hoogly river Ecosystem at Naihati (W.B.) Environ and Ecology, 10(1): 144–150.

Ghazali, Al. R.M. and Jairwal, F.S. (1988): "Antibiotic Resistance among pollution Indicator bacteria isolated from al–khair River Baghdad". Wat. Res. Vol. 22. No. 5: pp. 641–644.

Ghosh, B.B. and Basu, A.K.(1976): "Observation on eustarine pollution of Hooghly by the effluents from a chemical factory complex at Rishra, West Bengal, India" Environ. Health. Vol. 10. No. 3: pp. 204–218.

Ghosh, B.B., Roy, P. and Gopal K. (1973). Survey and characterisation of waste waters discharged into Hooghly estuary. J.Inland Fish Soc. India. 5: 82.

Glohmann, A. and Wassers, H.D. (1976). Hardness of drinking water and Public Health. Proceeding of the European Scientific colloquxm, Luxembourg: 129.

Gokhale, S. Kapadnis, B.P. and Patil, S.F. (1992). Impact of paper and pulp industry effluent on environment. NIE 1. (10): 6–11.

Gomati, V. (1992). Effect of pulp and paper mill effluent on germination of trac crops. Indian J. Environ. Health 1(4): 326–328.

Gondave, E.S. (1985). Supply of Chloride in ground water 2nd world congress on Engg. and Environment. 7–9, New Delhi.

Gonzalves, E.A. and Joshi, D.B. (1985). The seasonal sucession of the algae in a tank of Bandra J. Bombay Nat. Hist. Soc. 46: 154–176.

Goswami, M. and Naik, M.L. (1989). Physico–chemical study of a fertilizer factory effluent. Indian J. Ecol., 16(2): 115–118.

Graves, W.C., Burton, D.T., Richardson, B. and Margrey, S.L. (1981). The interaction of treated bleach kraft effluent and DO. Concentration on the survival of the development stage of sheep shead minnow cyprinodom variegatus; Water Res., 15: 1005.

Greenfield, R.E. (1925). Comparison of chemical and bacteriological exam. made on the ulionis river during a season of load and season of high water 1923–24. Ulionis slute water. Surg. Bull. 10: 9–33.

Griffth, R.S. (1955). Analysis of plankton yields of relation certain physical and chemical factors of lake Michigan Ecology, 36(4): 543–552.

Guantilaka, A. (1987). Nutrient cycling in tropical Asian Reservoirs: Some important aspects with special reference to Parakrama Samudra, Srilanka, In De Silva, S.S. (ed.) Reservoir Fishery management and development in Asia, IDRC : 33–46.

Gupta, M.P. and Battacharya, P.K. (1985). Studied on colour removal from bleach plant effluent of a kraft mill. J. Chemi. Tech. Biotech., 35(B): 25–32.

Gurnham, C. Fred. (1955): Principles of Industrial waste treatment, John wiley & Sons, Inc., New York. pp. 2–3.

Gutenmann, W.H., Bache, C.A., Youngs, W.D. and Lisk, D.J. (1976). Selenium in fly ash. Science, 191, 966–967.

Hammer, M.J. (1986). Water and waste water Technology. S.I. version. John Wiley & sons, New York: 1–536.

Hanumanulu, V. and Subrahmanyam, P.V.R. (1996). Economic of waste water treatment in small paper mill. IPPTA 14: 127–144.

Harned, D.A. (1982). Water quality of Nause river north Carolina variability. Pollution land and long term, water supply Paper 21–85.

Harris, Steven E. (1977): Intermittent sand filteration for upgrading waste stabilization pond effluents, J.W.P.C.F., 49(1): 83.

Hart, W.B., Doudroff, P. and Green Bank, J. (1948). The evaluation of the toxicity of industrial wastes, Chemical and other substances to fresh water fishes. Atlantic Refining Co., Philadelphia, 317–331.

Hawkes, H.A. (1975). Studies in Ecology, volume – 2. River Ecology, Oxford, 312–374.

Holden, J.M. and Green, J. (1960). Hydrology and plankton of the river Sokoto.J.Anim. Ecol., 29(1): 65–84.

Hood, D.W. (1966): "In, Fairbridge R.W. (ed). The Encyclopaedia of oceanography. van Nostrand, N. York.

Horne, A.P. and Bennison, L.G.(1987): "A laboratory stream design for biological Research". Water Res. Vol. 21. No. 12: pp. 1577–1589.

Hosetti, B.B., Patil, H.S., Rodgi, S.S. and Gaddad, S.M. (1985). Effect of detention period on biochemical activitiies of sewage stabilization ponds: A laboratory study. J. Environ. Biol. 6(1): 1–6.

Husmani, S.P. and Bharti, S.G. (1980). Limnologic studies in ponds and lakes of Dharwar Comparative phytoplankton Ecology of four water bodies, Phyco. 19(1): 27–43.

Hussain, S. (1987). Water supply and sanitary engineering. Oxford and IBH Publishing Co. Pvt.Ltd., New Delhi.

Hutchinson, G.E. (1967). A treatise on Limnology, introduction to lake biology and the Limnoplankton, 2, New York: John Willey and Sons, Inc., 1–1115.

Hynes, H.B.N. (1970). The Ecology of running waters. Liverpool University Press, Liverpool, 4th impression 1–585.

Hynning, Per–Ake (1996): "Separation, identification and quantification of components of Industrial effluents with Bio–concenteration potential". Wat. Res. Vol. 30 No. 5 pp. 1103–1108.

IAWPRC (1985): "Held a international singapore conference on Industrial water Technology, Treatment and Reuse of waste water." Wat. Res. Vol. 19 No. 4 pp. 543–545.

ICAR (1967): "Reports of the special committee on harmful effects of pesticides." I.C.A.R. New Delhi.

Ingram, W.J. (1989). In : Encyclopaedia of Environmental Science, Mc-Graw Hill Publication, II ed., pp. 663–665.

ISI, 1974. Tolerance limits for inland surface water for raw water used for Public water supplies and bathing ghats. New Delhi. Indian Standards Institution.

Jagals, P., Lues, J.F.R. (1996). The efficiency of a combined waste stabilization pond/maturation pond system to sanitise waste water intended for recreational reuse. Water science and Techol. 33(7): 117–124.

Jain, C.K. and Bhatia, K.K.S. (1996). Characterization of waste disposal and their impact on the water quaity of river Kali. Indian J. of Environ. protection. 17(4): 40–4.

Jain, S.K., Vasudevan, P. and Jha, N.K. (1990). Azolla pinnata R.Br. and Lemna minor L. for removal of lead and zinc from polluted water. Wat. Res. Vol 24(2): 177–183.

Jenke, R. Dennis and Diebold, E. Frank (1985): "Computer simulation of an Industrial waste water treatment process." Wat. Res. Vol. 19 No. 6 pp. 719–724.

Jindal, R. and Vasisht, H.S. (1985). Ecology of fresh water pond at Nabha Punjab (India) Zoological Orientalis (18): 74–85.

John, R., Holum (1977). Topics and term in environmental problems in New York. John Wiley and Sons, New York (1977 Ed): 655.

John. V. and Alexander, K.M. (1968). A preliminary report on the hydrobiology of Beypore river for the year 1964. Hydrobiologia, 31:492–496.

Jolley and Jolley. *et al.*(1975–85): "Environmental Hazards due to excessive chlorincation waste water treatment." Wat. Res. Vol. 2 No. 6 pp. 233–241.

Joshi, B.D. Bisht, R.C.S. and Semval, V.P. (1995). Primary productivity in western Ganga Canal at Haridwar India. J. Ecol. 22 (2) : 123–126.

Joshi, C.B. (1990). Hydro–Biological Profile of river Sutlej in its middle stretch in western Himalayan, U.P. 1 Zoal. 16 (2) : 9–103.

Joy, C.M. (1989): "Growth response of phytoplankton Exposed to industrial effluents in river Priyar." Ph.D. thesis Cochin University of Science and Technology, India.

Juanico, M. and Shelef, G. (1994). Design, operation and performance of stabilization reservoirs for waste water irrigation in Israel. Wat. res. Vol. 28 : 175–186.

Jusiak, P.M. and Kosinska, E. (1984): "Removal of Nitrogen from Industrial waste water with the use of algal rotating disk and denitrification packed bed reactor." Wat. Res. Vol. 18 No. 9. pp. 1077–1082.

Kamath, P.R. (1985): "Radioactive pollution in the quality in the Ganges." (Water pollution and management journal).

Kannan, P.K. and Oblisami, G. (1990). Influence of paper mill effluent irrigation on soil enzyem activity. Soil Biol. Biochemi., 22(7): 293–293.

Kant, K.A. and Siddiqui, A.Qs (1971). Diurnal variations in the Pond Moat at Aligarh, J. inland fish. soc. India, 2: 146–154.

Kant, S. and Vohra, S. (1993). Role of algae as primary producer in fish production in Kashmir lakes. Advances in Limnology, Ed. H.R. Singh, Narendra Publishing House, Delhi : 79–86.

Kasuhik, P. 1983. Ecological and Anatomical Marvels of the Himalayan Orchids. Today and Tomorrow's Printers and Publishers, New Delhi, pp. viii + 123 + plates 71.

Kaushik, P. (1988). Indigenous Medicinal plants including Microbes and Fungi. Today and Tommorow's Printers and Publisher's New Delhi. pp vii + 243.

Kaushik, P. (1996). Introductory Microbiology. Emkay Publications. Delhi. pp viii + 344.

Kaushik, P. and Ghanaksh, A. (1998). Water quality of Bathing Ghats of Ganga at Haridwar. Abstract No. 45, pp. 29, Souvenir and Abstracts of National symposium on Faunistic Bio–Diversity, Environmental monitoring and Biotechnology. Department of Zoology and Environmental Sciences, G.K.V. Haridwar.

Khanna, D.R. (1993). Ecology and pollution of Ganga River, Ashish Publishing House, New Delhi.

Khanna, D.R. Badola, S.P. Singh, H.R. and Dohriyal, K. Anoop (1992). Observation on seasonal trends in Diamotic Diversity in the River Ganga at Sapt–Sarovar, Haridwar. Recent Research in Cold water fisheries, ed. by K.L. Sehgal, Today and Tomorrow's Printer. New Delhi. pp. 99–107.

Khanna, D.R., Badola, S.P. (1993). Plankton Ecology of the River Ganga at Chandighat, Haridwar, Advances in limnology, ed. H.R. Singh, 171–174.

Khare, V.S. and Sharma, E.C. (1979): "Ecology of Solah sugar pond." Geobios. Vol. 3 No. 2: pp. 505–519.

Kim, R. Byung, Anderson, J. Stanley and Zemla, F. Jerome(1990): "Effect of biological treatment on COD absorption." Wat. Res. Vol. 24 No. 4 pp. 457–491.

Kobayashi, K. (1977). Metabolism of penta chlorophenol in fish cited in pentachlorophenoledifed, Plenum, New York pp. 89.

Koe, Lawrence C.C. and Tan, N.C. (1990). Odour generation potential of waste waters. Wat. Res. Vol. 24(12) : 1453–1458.

Kolev, N., Semkov, Kr. and Darakchiev, R. (1996): "Butyl Acetate and Butanol stripping from waste water in Antibiotic production." wat. Res. Vol. 30 No. 5 pp. 1312–1315.

Kolkwitz, R and Marsson (1908): "Okologie oder Pflanzlichen Saprobein." Ber. Deutsch Bot. Ges. Vol. 26. pp. 505–519.

Koschel, R. and Mothes, G. (1976). Change of the biological productivity in a lake system through the in How of cooling water from a nuclear power plant. International symposium on eutrophican and Rehabilitation of surface waters, 3, 209–223. Karl–Marx–St.

Kshirsagar, S.R. (1988): Sewerage and sewage Treatment, 7th ed. pp. 240–249.

Kudesia, V.P. (1980). Water Pollution, Pragati Prakashan, Meerut : 1–300.

Kumar, A. (1995). Studies on pollution in river Majurukshi in South Bihar. Indian J. Env. Poll. 2 (1) : 21–26.

Kumar, H.D. (1965): "Effect on certain tonic chemicals and Mutagens on the growth of the Blue Green Algae." Can. J. Bot. Vol. 43 : pp. 1523–1532.

Kumar, H.D., Bisaria, G.P., Bhandari, L.M., Sharma, V. and Rana, B.G. (1974): "Ecological studies on algae isolated from effluents of an oil refinery and fertilizer factory and a Brewery." Indian, J. Environ. Health. vol. 16 : pp. 379–410.

Kumar, J.I.,Nirmal, R and Rana, B.C. (1991). Physico–chemical proporties of certain industrial effluents of Central Gujrat J. Ind. Poll. Cont. 7(1): 17–24.

Lackey, J.S. (1942): "The effects of distillary wastes and water on the microscopic flora and fauna of a small creek." Public Health Repts. Vol. 57 : pp. 253–260.

Lakshminaryana, J.S.S. (1965). Studies of the phytoplankton of the river Ganges Varanasi. Hydrobiologia. 25: 119.

Lau, P.S., Jam, N.F.y. and Wong, Y.S. (1995). Effect of algal density on nutrient removal from primary settled waste water. Env. Pollution, 89: 59–66.

Laxen, H.P. Duncan and Harrison, M. Roy (1983): "Physico–chemical speciation of selected metals in the treated effluent of Lead– Acid battery manufacture and in the receiving river." Water. Res. Vol. 17: pp. 71–80.

Lehr, J.H., Gass, T.E., Pettyjohn, W.A. and Damaree, J. (1980). Domestic water treatment. McGraw–Hill Book Co.

Lens, P.N., de Poorter, M.P., Croneberg, C.C. and Verstraete, W.H. (1995). Sulfate reducing and methane producing bacteria in aerobic waste water treatment systems. Wat. Res. Vol 29(3): 871–880.

Liran, A., Juanico, M. and Shelef, G. (1994). Coliform removal in a stabilization reserviour for waste water irrigation in Israel. Wat. Res. 28(6): 1305–1314.

Livingston, R.J. (1978). Respiration growth and food conservation efficiency of pin fish (Langodom rhomboides) exposed to sublethal concentration of bleached kraft mill effluent ; Environ. Poll., 17:207.

Madhwal, B.P., Chopra, A.K. and Singh, H.R. (1983). Diurnal fluctuactions in physico–chemical parameters of ther river Yamuna from the Garhwal Himalaya. Uttar Pradesh J. Zool. 3: 157–158.

Madoni, P. Davoli, D. and Chierici, E. (1993). Comparative analysis of the activated sludge microfauna in several sewage treatment works, Wat. Res. Vol 27(9): 1485–1491.

Mahaputra, P.K., Patnaik, L.N. and Mishra, Gadadhar (1993). Water pollution due to textile mill. J. Indl. poll. cont., 9(2): 51–63.

Mahen Mani, K. Raja Kumari, J. and Pongallappn, S. (1991). Effect of Tannery effluent on the quality of an irrigation canal water 1991. J. Ind. Poll. Count, 7 (2) : 87–91.

Malanchuck, J.L. and Gruendling, G.K. (1973): "Toxicity of lead Nitrate on Algae." Jour. of water air, soil pollution. Vol. 2 : pp. 180–190.

Mara, D.D. (1974). Bacteriology for Sanitory Engineers, Churchill Livingston, Edinburgh.

Marssor., *et al.*. (1908): "Okologie der pflanzlichen saprobein." Ber Deutsch Bot. Ges. Vol. 26 : pp. 505–519.

Mather, K. (1973). Statistical Analysis in Biology. Chapman and Hall Ltd., London.

Mathur, R.P. (1993). Inc. Water and Waste Water Testing (A Laboratory Manual), Nem Chand & Bros Publ. Roorkee.

Mazhar, M.D. and Kapoor, C.P. (1992). Innologican studies on Doramin river at Barily (U.P,) J. fresh water Biol 4 (2) : 155–158.

Mckay, Gordon and Bino, J. Murad (1988): "Absorption of pollutantes from waste water onto activated carbon based on Eternal Mass transfer and pore diffusion". Wat. Res. Vol. 22 No. 3 pp. 279–286.

Mechalas, B.J., Hekimian, K.K., Schinazi, L.A. and Dudley, R.H. (1972). An investigation on to recreational water quality. Water Quality Data Book, 4, US Environmental Protection Agency.

Mehrotra, S.N .and Jhingran, A.G. (1986). Evaluation of limnologic characteristics and tropic status of Gularya reservoir India. Proc, Nat, Acad Sci India B56(3): 204–210.

Metcalf and Eddy (1972). "Waste Water Engineering, McGraw–Hill Inc (1972)", New York.

Michael, R.G. (1968). Studies on the Zooplankton of tropical fish pond. Hydrobiologia, 32: 47–68.

Millington, L.A. and Adams, N. (1988): "The influence of growth medium composition on the toxicity of chemicals to Algae". Wat. Res. Vol. 22 No. 12 pp. 1593–1597.

Mishra, B.N. and Behra, B.K. (1982). Analysis of the industrial effluent on growth and development of rice seedling. Env. Res., 28: 10–20.

Mishra, R. and Saksena, D.N. (1990). Zooplanktonic fauna of lentic waters at Bhind (M.P.) Geobios New reports 8(2): 115–118.

Mistry, Khursheed J. and David, M. Himmel blau (1975). Stochastic analysis of trickling filter, J. Env. Div. Proc. American Society of Civil Engineers: 101.

Mitra, A.K. (1982). Chemical characteristics of surface water at selected ganging section in the river Godavari, Krishna and Tungbhadra. Indian J. of environ. Health 24(2): 165–174.

Mittar, D., Khanna, P.K., Marwaha, S.S. and Kennedy, J.F. (1992). Bio–bleaching of pulp and paper mill effluent by phanerochaite chrysosporium. J. Chem. Tech. Biotech., 53: 81–92.

Mohan, C. Balani and Sarkar, H.L. (1965). Some observation on the pollution of Yamuna river at Okhla water works in takes, Delhi. Indian J. Environ, Hlth. 7: 84–86.

Moller, J. and Chalkins, J. (1980). Bacteriological agents in waste water lagoons and logoons desigr J. Wat. Poll. Control Fed. 52(10): 2442–2451.

Monawar, M. (1970). Limnologic studies on fresh water ponds of Hyderabad India. The biotope, Hydrobiologia 35, (1): 127–162.

Moore, J.E. and Love, R.J. (1977). Effect of pulp and paper mill effluent on the productivity periphyton. J. Fish Res., Bd. Can., 34: 8156.

Morgan, G.B. and Lackey, J.B. (1985). BOD determinations in waters containing chelated copper and chromium. Sewage Ind. Wastes, 30, 283–286.

Moss, Brian (1982). Ecology of fresh waters. Black well scientific publications, Oxford, 1–332.

Motwani, M.P. Bohersi, S.M. and Karam Chandani, S.J. (1986). Some observation on the pollution of the river some by the factory effluents of Rolitas industries at Dalmia Nagar (Bihar). Indian J. fish 3 : 334–367.

Motwani, M.P., Banerjee, S. and Karanchandani, S.J. (1956). Observation on pollution of the river sone by factory effluents at the Rohtas industry, Dalmia Nagar, Bihar. Indian J. Fish.3: 334–367.

MPPCB (1994). Utilization of fly ash in agriculture analysis of samples, Madhya Pradesh Pollution Control Board, Bhopal, Madhya Pradesh.

Munshi, J.S.D. and Rai, D.N. (1979). Observations on diurnal changes of same physico chemical factors of three tropical swamps of Darabhanga (W.Bihar) India Comp. physical Ecol., 4 (2): 50–85.

Nampoothery, M.K. and Sashidhram, K.H. (1976). Pollution control in pulp and paper mill. IPPTA, (9): 7–10.

Nandan, S.N. (1985). Eutrophication in Vishwamitri river following through Baroda city (India). Geobio. Science, 12(2): 60–62.

Narayanaswamy, M.S. (1982). River water quality for bathing purposes. Journal of the Institution of Engineers. 62: 59–61.

Nasar, A.K. (1977). Investigation on the seasonal periodicity of Zooplankton in fresh water pond in Bhagalpur India, Acta Hydrochem, Hydrobiol, 5: 577–584.

Nayar, K.G. (1970). Studies on the rotifers population of two ponds at pilani Rajasthan J. Zool. Sco. India, 22(182): 21–34.

Naytiyal, P. (1990). Ecology of the Ganga river system in the upland of Himalayan Environment. Resource and Development. 2(1): 69–73.

Needham, J.G. and Needham, P.R. (1972). A guide to the study of fresh water Biology. Holden–Day I.N.C. San Francisco, Colif., 94(3): 1–108.

Neelay, V.R. and Dhondizal, L.P. (1985). Observation on the possibility of using industrial effluent water for raising forest plantation. J. Trop. Forestry, (2): 131–139.

Nikolsky, G.V. (1963). The ecology of fishes. Academic Press, London, 1–352.

Okari, A., Lonn, B.E. and Virtanen, E. (1983). Toxicological effect of dehydroubutic acid (DHAA) in the trough solmogainedneri Richardson in fresh water. Water. Res., (17): 665–671.

Olach, Jozsef, Cserhati, Tibor & Szejtl, Jozsef (1988): "B–yclodextrin enhanced biological detoxification of Industrial waste waters". Wat. Res. Vol. 22 No. 11 pp. 1345–1351.

Oliff, W.O. 1959. Hydrobiological studies on the Tugela river system. Part I. The main Tugela River. Hydrbiologica, 14 : 281–370.

Olin, Robert, G., Andres, Ahlbom and Norell, E. (1987). Occupational study American. J. Ind. medical 11(6): 615–626.

Osborne, L. Lewis, and Davis, W. Ronald (1987): "The effects of a chlorinated discharge and a thermal outfall on the structure and composition of the Aquatic Macro–invertebrate in the sheep river, Alberta, Canada". Wat. Res. Vol. 921 No. 8 pp. 913–921.

Otmar, O. Heuvelen, Oseon. Willis, Van and Vennues, John. W. (1949). Combined Industrial and Domestic Waste Treatment in waste stabilization lagoons, J.W.P.C.F., 40(2): 214–222.

Page, A.L., Elseewi, A.A. and Straugham, I.R. (1978). Physical and chemical properties of fly ash from coal fired power plants with reference to environmental impacts. Residue Rev., 71, 83–120.

Paka, Swarnalatha and Rao, Narsing, A. (1997). Inter–relationships of physico–chemical factors of a pond, J. Env. Biol 18(1): 67–72.

Paliwal, K.V. (1983). Pollution of surface and ground water 30–60 in C.K. Varshney. Water pollution and management, Wiley Eastern limited, New Delhi.

Palmer, C.M. (1980). "Algae and water pollution".

Palmer, C.M. (1969). A composite rating of algae tolerating organic polluttion: J. Phycol. (5): pp. 78–82.

Pandey, Anuradha, and Garg, S.L. (1992). Recycling of effluent water and its utilization. Pollution control aspect in pulp and paper industries; Indian J. Environ. Prot., 12(9): 659–663.

Pandey, B.N., Kumar, K. Lal, A and Das, P.K.L. (1993). A Preliminary study on the physico–chemical quality of water of the river Koshi in J. Ecobial 5: 237–339.

Pant, M.C., Gupta, P.K., Pandey, J., Sharma, P.C. and Sharma, A.P. (1981). Aspect of water pollution in lake Nainital U.P. India. Environmental Conservation 8(2): 113–115.

Patric, R. (1949): "A proposed biological measure of stream conditions, based on a survey of the connestoga Basin Lancaster country Pennsylvania". Proc. Acad. Nat. Sci. Philadelphia. Vol. 101 : pp. 277–341.

Patric, R. (1950): "Biological measure of Stream conditions". Verth. Int. Sew. Indust. wastes Vol. 22 : pp. 926–938.

Patric, R. (1953): "Aquatic organisms as an acid in solving waste disposal problems". Jor. of Sewage & Industrial waste. Vol. 25 : pp. 210–216.

Pearson, H., Mara, D., Mills, S. and Smallman, D. (1987). Physico–chemical parameters influencing faecal bacteria survival in waste stabilization ponds, Water Sci. Technol. 19(12): 145–152.

Pelczar Michael J. Jr., Chan E.C.S. and Krieg Noel R. (1995): Microbiology of domestic water and waste water, "Microbiology", 15th Ed. 1995, pp. 607.

Pelczar, M.J., Chan, E.C.S. and Krieg, N.R. (1993). Microbiology: Concepts and applications. McGraw–Hill Inc. New York : 1–896.

Peter, A.K. (1974). Sources and classification of water pollutants. Industrial pollution. Van Nostrand Reinhold Company.

Prasad, B.V. and Chandra, Ramesh, P. (1997). Ground water quality in industrial zone. A case study. Poll. Res 16(2): 105–107.

Prasad, D.Y. (1990). Primary productivity and energy flow in upper lake Bhopal. Indian J. Env. Heth., 32, 2 : 132–139.

Probst, J.L. (1986). Dissolved and suspended matter transported by the Gira. River (France): Mechanical and chemical erosion rates in a caloriosu molanebasin. J. Des. Science Hydrobiologious 31: 61–79.

Quadri, M.Y. and Yousuf, A.R. (1978). Seasonal variations in the Physico–chemical factors of a subtropical lake of Kashmir, J. inland Fish. sec. India, 10: 89–96.

Quasim, S.Z. and Siddique, R.M. (1969). Pollution of river Kali caused by effluent of industrial waste water; Current Science. (27): 310–311.

Rai, H. (1964). The bacteriological studies on the Yamuna river (Delhi, India). Surc. paper from Inst. Chem. Tech. Prague,: 319–336.

Rai, H. (1974). Limnological studies on the river Yamuna at Delhi, Part I: Relation between the chemistry and the state of pollution in the river Yamuna. Arch. Hydrobiol., 73: 369–393.

Rai, L.C. (1978). Ecological studies on algal communities of the Ganga river at Varanasi. Indian J. Ecology. 5: 1–6.

Raina, H.S. Shah, A.R. and Ahmed S.R. (1984). Pollution studies on river Jhelum 1. An asessment of water quality, Indian J. Env. Health : 26 : 187–210.

Raina, V., Shah, A.R. and Shakto, A.R. (1984). Pollution studies on water quality. Indian J. Environ Hlth. 26: 187–201.

Rajczyk, Janosz M. (1993): "Fermentation of food industry waste water under dynamic condition in an anaerobic biofilter up flow". Wat. Res. Vol. 27 No. 7 : pp. 1257–1262.

Ramesh, Y. and Ramanujam, M.P. (1991). Effect of drainage effluents on seedling growth of Ground Nut, Ad. Plant. Sci., 4(2): 425–429.

Rana, B.C. and Palria, S. (1987). Pollution abatement of certain waste waters using biological and chemical methods, Ind. J.Ecol., 14(1): 1–6.

Ranjannan, G. and Oblisami, G. (1979). Effect of factor effluent on soil and crop plants. Indian J. Envron. Health. (21): 120–130.

Rao, D.S. and Govind, B.V. (1966). Hydrology of Tungbhandra reservoir Indian J. Fish, 2 (i): 321–344.

Rao, J.C. and Rani, Sudha K. (1985): "Effect of Mercuric chloride on photosynthesis and respiration of scenedesums incrassauflus". Geobios. vol. 12. : pp. 82–83.

Rao, N.G. and Durve, V.S. (1989). Cultural Eutrophication of the lake Ranasagar Udaipur Rajasthan, India. J. Environ. Biol. 10(2), 127–134.

Rao, S.S., Kwiatkowski, R.E. and Turkovie (1979). Distribution of bacteria and chlorophyll 'a' at a nearshore station in lake antario. Hydrobiologia, 66 : 33–39.

Ratsak, C.H., Maarsen, K.A. and Kooijman, A.L.M. (1996). Effects of protozoa on carbon minerlization in activated sludge. Wat. Res. 30: 1–12.

Rawl, N.C. (1978). Quality of river water of India, Reprinted from proceedings fourty seven research session of the CBIP: Hubli Dharwar Karnataka, II: 139–160.

Ray, P. and David, A. (1966). Effects of industrial wastes and sewage upon the chemical and biological composition and fisheries of the river Ganga at Kanpur, Ind. J. Env. Hith., 8: 307–338.

Ray, P., Singh, S.B. and Sehgal, K.L. (1966). A study of some aspects of the river Ganga and Yamuna at Allahabad (U.P.) in 1958–59. Proc. wat. Acad. Sci., India 36B (3): 238–272.

Reddy, Manika, P. and Venkateswarlu, V. (1987). Assessment of water quality and pollution in the river Tungabhadra near Kurnool (A.P.) J. Environ. Biology. 8(2): 109–199.

Richter, L.A., Volkov, E.P., Pokrovsky, V.N. (1984). Thermal power plants and Environmental control., Mir publication Moscow.

Riedel, K., Lange, K.P. and Scheller, F. (1990): "A microbial sensor for BOD". wat. Res. vol. 24 No. 7 : pp. 883–887.

Roald, B.O. (1977). Effect of sublethal concentration on lignin sulphonates on growth intestinal flora and some digestive enzyme of rainbow trout. Agriculture. 12: 327–332.

Robert, D., Barshied and Hussan (1974): "Physical and chemical treatment of septic tank effluent". Jour. WPCF Vol. 46. No.10 : pp. 2347–54.

Robinson, J.M. and Kaller, E.C., 1976. A comparison of the water characteristics of four northern west Virginia rivers. Jr. Proc.w.va.Acad.Sci., 48(1): 1–67.

Romney E.M., Wallace A. and Alexander G.V. (1977). Boron investigation in relation to a coal burning power plant. Communications in soil Science and plant Analysis. 8, 803–807.

Rydin, E. (1996). Experimental studies simulating potential phosphorus release from municipal sewage sludge deposits, Wat. Res. Vol 30(7) : 1695–1701.

S. Subramani and Maha Devan, A. 1995. Water quality of river (River Suruliyar in Tamil Nadu, Pall Res. 14(1) : 73–82.

Sacramento, C.A. (1963). Water quality criteria 2nd ed. California state water quality control board, 263.

Salim, Radi (1983): "Absorption of lead on the suspended particles of River water". Wat. Res. Vol. 17 No. 4 : pp. 423–429.

Sangu, R.P.S. and Sharma, K.D. (1985). Studies on water pollution on Yamuna river at Agra, Indian J. 11th, 27(3) : 257–261.

Sapru, R.K. 1989. Water pollution management (A case of *Chansigarh*) Environment pollution and management. Wiley Eastern, New Delhi. pp. 303–318.

Saqqar, M. and Pescod, M. (1990). Microbiological performance of multistage stabilization ponds for effluent use in agriculture. Wat. Sci. Technol. 23(7–9): 1517–1524.

Sastry, C.A. (1977). Treatment of waste water from small paper mill without soda recovery. A case study. Indian J. Environ. Health (19): 346.

Sastry, C.A., Alagassamy and Kothandaraman (1974). Treatment of pulp and paper board mill waste a case study. Indian J. Environ. Health. 16(4): 317–335.

Saxena, K.L., Chakraborty, R.N., Khan, A.Q. and Chattopadhya, S.N. (1966). Pollution studies of the river Ganga near Kanpur. Indian J. Environ. Hlth. 8: 270–285.

Saxena, R. C. (1983). Effect of physico–chemical factor on Benthic fauna of Bhagirathi river, Garhwal Himalaya, Indian, J. Ecol. 13 (i) : 133–137.

Scheimer, F. (1983). Limnology of Parakrama Samudra, Sri Lanka a case study of an ancient man make lake in the tropics. Development in Hydrobiology. 12 : 236.

Schineder, H.A.I., Soraes, A. and Rech, R.L. (1995): "Primary treatment of a soyabean protein bearing effluents by dissolved air floatation and by Sedimentation". Wat. Res. Vol. 29 No. 1 : pp. 69–75.

Scrimgeour, G.J. (1989). Effect of bleached kraft mill effluent on macroinvertebrate and fish population in weed beeds in a New Zealand hydroelectric lake N.Z.I., Fresh water Res. 23,(3): 373–379.

Sekerka, I and Lechner, J.F. (1975). Simultaneously determination of total, non–carbonate and carbonate water hardness by total potentionmetry. Talanta. 22 : 459.

Sengupta, A. and Chaudari, S. (1989). Self purification of untreated metropolitan waste water of Calcutta under saline ecosystem, Ind. J. Ecol., 16(1): 180–182.

Sharma S.R., Rathore H.S., Ahmed S.R. (1983): "A new specific spot test for the detection of Malathionin water". Wat. Res. Vol. 17 No. 4 : pp. 471–477.

Sharma, A., Naik, M.L., and Sharma, A. (1990). Physico–chemical characteristics of a steel plant waste water and its effects on soil and plant characteristics, Ind. J. Ecol. 17(1): 9–12.

Sharma, B.K. and Kaur, H. Inc.: Environmental Chemistry, Goel Publishing House, Meerut (1994–95).

Sharma, C.R., Juyal, C.P., Gusain, O.P., Rawat, M.S. (1993). First report of the limnology: Abiotic profiole of a Garwhal Himalayan lake, Deoriatal. Advances in Limnology, Ed H.R. Singh Narendra Publishing House, Delhi : 87–92.

Sharma, D.R.R., Kumarswami, N. and Venkates, Rao M.V. (1989). Assessment of industrial water in Sivakasi (south India); Indian J. Environ. Health 33(3): 330–335.

Sharma, R.C. Gusain, O.P. Juyal, C.P. and Rawat, M.S. (1989). Environmental studies and management of lakes in Garwhal Himalaya : Containts and prospects JOHSARD.

Shastry, C.A., Khare, G.K. and Rao, A.V. (1972): "Water pollution problem in M.P". Indian Jour. Environ. Health Vol. 14 : pp. 297–309.

Shende, G.B. and B.B. Sunderson (1978): "Nord Seminar on the development of complementary use of mineral fertilizers and organic Material in India". Ministry of Agriculture, New Delhi.

Shrivastava, S.K., Bembi., R., Singh, A.K. and Sharma, A. (1990). Physico–chemical studies on the characteristics and disposal problem of small and large pulp and paper mill effluent. Indian J. Environ. Prot., 10(6): 441.

Shukla, S.C., Tripathi, B.D., Rajanikant, Kumari, D.V. and Pandey, V.S. (1989). Physico–chemical and biological characteristics of river Ganga have been studies from Mirzapur to Ballai. Indian J. Environ. Hlth. 31: 218–227.

Shuval, H.I. (1974). Disinfection of waste water for agricultural utilisation, Prog. Wat. Technol., Vol. 7: 857–867.

Shyam Sunder 1996. Biotic communities of Kumaon, Himalaya river, U.P. J. zoal 10(1) : 39–45.

Siegal, S.M. and Eshleman, A. (1975): "Chlorine bioaches: A significant long term source of mercury pollution". Jour. of water, air, soil pollution Vol. 4 : pp. 355–365.

Sierra, Alvareze. (1991). The anaerobic biodegradability and methanoganic toxicity of pumping waste water. Water sci. Technology (24): 113–125.

Sikander, M and Tripathi, B.D. (1984). Physico–chemical characteristics of Ganga water at Varanasi. National Env. Consery Assoon. 53–61.

Sikander, M. (1987). Ecology of river Ganga in Varanasi with special reference to pollution. Ph.D. Thesis, Banaras Hindu University, Varanasi.

Simsima, G.V., Chesters G. and Andren, A.Q. (1987). Effect of ash disposal ponds on ground water quality at coal fired power plant, Wat. Res, 21, 417–426.

Singh Room and Mahajan, I. (1987). Phytoplankton and water chemistry of Renuka lakes, Himachal Pradesh. Ind. J. Ecol 14(2): 273–277.

Singh, A. and Wadhwani, K. (1987). Fungi in relation to physico–chemical factors of tube well waters. J. Env. Bio., 8(4): 347–351.

Singh, B.S. and Sahai, K. (1986). The primary production of phytoplankton of a tropical shallow water body of Gorakhpur Tropical Ecol. 27: 33–39.

Singh, G.S. and Singh, A.S. (1994). variation and correlation of DO with effluent quantity and stage of river Ganga at Varanasi (India). Indian J. Environ. Health 36(2): 79–83.

Singh, H.P. Choudhry, M. and Kalckaur, V. (1982). Seasonal and Diurnal changes in physico–chemical features of the river Brahamputra at Gauhati, Ind. J. Zoot. 2(82): 77–84.

Singh, H.R. (1986). Pollution study of the upper Ganga and its tributaries, A progress report. Garhwal University, Srinagar.

Singh, M.P. (1988). Effect of waste disposal on water quality of river Damodar in Bihar. In Ecology and Pollution of Indian river. (Ed. By R.K. Trivedi). Ashish publishing House, New Delhi.

Singh, P.K (1975): "Fertilizer tolerance of Blue–green algae and their effect of Heterocyst differentiation". Phykos vol. 14 : pp. 81–88.

Singh, T.N. and Singh, S.N. (1995). Impact of river Varuna on Ganga river water quality at Varanasi Indian J. Environ. Health 27(4): 272–277.

Sinha, S.N. and Banerjee, R.D. (1987). Pollution of the Ganga water near Palta, West Bangal. Environment and Ecology 5 : 137–143.

Somasekhar, R.K., Gowda, M.G. and Singh, K.P. (1984). Effect of industrial effluent on crop plants; Indian J. Environ. Health (26): 136–146.

Sreenivasan, A. (1966). Limnology of tropical impoundments I. Hydrobiological features and fish production in stanley reservoir, Mettur Dam Int. Rev. ges. Hydrobiol. 51(2). 295–306.

Sreenivasan, A. (1970). Limnology of tropical impoundments, a comparative study of the major reservoirs in Madras States (India). Hydrobiol. 36(3–4) : 443–469.

Sreenivasan, A. (1984). Influence of stocking on fish production in reservoirs in India. FAO fisheries, Report No.312, FAO, Rome.

Srivastava, K.S. and Jain, K.C. (1985): "Estimation of sulphate ions in water and industrial wastes using a solid state membrane electrode". Wat. Res. Vol. 19 No. 1 : pp. 53–56.

Srivastava, R.K. and Gupta, S.K. and Nigam, K.D.P. and Vasudevan, P. (1994). Treatment of chromium and nickel in waste waters by using aquatic plants, Wat. Res., 28(7): 1631–1638.

Subba Rao, N., Ravi Prakash, and Krishan, G. (1988). Ground water pollution by industrial effluent in Visakhapatnam area; Indian J. Environ. Prot., 10(5): 336–341.

Subhas Chandra, Gaddad, M. and Shiva, Rodgi, S.S. (1987). Effect of temperature on the growth and biochemical activities of *Pseudomonas aeruginosa*, isolated from a stabilization pond. J. Env. Biol. 8(2): 143–149.

Subrahmanyam, P.V.R. (1975). Colour removal from kraft pulp mill waste water stater of art. IPPTA souvenin, 58: 108–144.

Subrahmanyan, P.V.R. (1990). Waste management in pulp and paper industry. J. Indian Assoc. Environ. Manag. 17: 79–94.

Sudhakar, G., Venkateshwar, V. (1989). Ecological balance in the river of Andhra Pradesh (India) Effect of paper mill effluent on river Tungabhadra and Godavari; INT J. Environ. Study, 34(1–2): 89–97.

Suess, J.M. (1982): "Biological, Bacteriological and Virological Examination of water for pollution control". Indian, J. of world Health organisation : pp. 531.

Swami, M.S.R., Kumar, Vijay, V. and Sai, C.S.T. (1987). Ground water contamination in IDA Jeedimetla Hyderabad by industrial effluent. J. Indl. poll. cont., 3(2): 73–80.

Swarup Krishan and Singh, S.R. (1979). Limnologic studies of Suraha lake (Ballia). J. Inland Fish Soc. India, 2(1): 22–23.

Syers, J.K. and Iskandar and Keeney, D.R. (1973). Distribution and background level of mercury in sediment cores from selected Wisconsin lakes: Water air and soil pollution, 2, pp. 105–118.

Symons, G.E. (1956): "Taste and odour control". Water and Sewage wks. Vol. 38. : pp. 307–310.

Talbot, P. and de la Noue, J. (1993). Tertiary treatment of waste water with Phormidium bohneri (Schmidle) under various light and temperature conditions. Wat. Res. 27(1): 153–159.

Tambe, M. (1986): "Biotechnology applied to distillery waste water treatment A new tech". Jour. CEW. Vol. 21. No. 4 : pp. 59–66.

Theis, T.L and Richter, R.O. (1979). Chemical speciation of heavy metals in power plants ash pond leachate. Environ. Sci. Technol., 13, 219–224.

Theis, T.L., Westrick, J.D., Hsu, C.K. and Marley, J.J. (1978). Field investigation of trace metals in ground water from fly ash disposal. J. Water Poll. Cont, Fed., 50, 2457–2469.

Thomas, J.F.J. (1953). In : Industrial waste water resource of Canada. Water survey report No. 1 scope procedure and interpretation of survey studies. Ottawa, Queen's printer.

Tiwari, T.N. and Manzoorali (1987). River pollution in Kathmandu Valley. Variation of water quality index, Indian J. Environ. Prot., 7: 347–351.

Townsend, W.N. and Gilham, E.W.F. (1975). In : Pulverised fuel ash as a medium for plant growth. The Ecology of Resource Degradation and Renewal (edn. Chadwick M.J. and Goodman G.T.). 287–304. the 15th symposium of the British Ecological Society, Blackwell, Oxford.

Tripathi, K.A. and Pandey, S.N. (1995) In: "A literary Survey on water pollution". Pub. Ashish Publishing House, New Delhi : pp. 1–320.

Trivedi, R.C. 1990. Pollution in Indian river, Ashish Publ. House, New Delhi.

Trivedi, R.K. and Goel, P.K. (1984). Chemical and Biological Methods for Water Pollution Studies. Environmental Publication, Karad, pp. 1–251.

Trousellier, M., Legendre, P. and Baleux, B. (1986). Modelling the evolution of bacterial densities in an eutrophic ecosystem (sewage lagoons), Microb. Ecol., 12: 355–374.

Turner, R.R., Lowry, P., Levin, M., Linberg, S.E. and Tamura, T. (1982). Leachability constituents of coal fly ash. EPRI. Palo Alto, California.

Ubom, H. Gregory, and Tsuchiya, Y. (1988): "Determination of Iodine in natural water by Jon chromatography". Wat. Res. Vol. 22 No. 11 : pp. 1455–1458.

Unadri, and Shah, G.M. (1984). Hydrobiological features of Hoassari typical wetland of Kashmir–1, Biotope, Indian J. Ecol. 2(2) : 203–206.

Vaidyanathan Sai R., "Poisoned River Every Where", The Tribune, Thursday, Nov. 7, 1996.

Velz, C.J. and Gannon, J.J. (1960). Forecasting heat loss in ponds and streams. J. Wat. Pollut. Control Fed. 32, 392–417.

Venkatesh, V., Bhupathiganesh, T. (1992). Application of parametric ratio in the evaluation of ground water pollution potential due to chemical industrial effluents. Indian J. Environ. Prot. 12(10): 757–759.

Venkateshwarlu, T. and Jayanti, T.V. (1968). Hydrobiological studies of the river Ganga and its branches. Hydrobiologi 9 : 46–70.

Venkateswarlu, V. (1981): "Algae as indicator of river water quality and pollution. Biological indicator and Indices on Environmental pollution". Cent. Bd. Pre.Cont. Water poll. OSM Univ. Hyderabad, India.

Verma, R.B. and Das, Keshav (1982). Ecology of the Bntler Palace Tank, Lucknow. Uttar Pradesh. J. Zool., 2 : 109–111.

Verma, S.R. and Shukla, G.R. (1969). Pollution in perennial stream khala by sugar factory effluent near Laksar (District Saharanpur) U.P.; Indian J. Environ. Health (11): 145–162.

Verma, S.R., Dalela, R.C., Tyagi, M.P. and Gupta, S.P. (1974). Studies on the characteristics and disposal problem of the industrial effluent with reference to ISI standard part I. Indian J. Environ. Health 16(4): 289–299.

Vijay Kumar, V. and Ra, P.L.K.M. (1994). Impact of effluent of cracker manufacturing units on ground water in Sivakashi. J. Indian Polll Cant. 10(2): 145–148.

Vijayam, K., and Vasugi, S.R. (1989). Sublethal effect of pulp and paper mill effluent on the Biochemistry of a fresh water Rasbara Daniconius. Indian J. Environ. Health (31) (1): 36–42.

Volz, C.J. 1947. Factor influencing self purification and their relation to pollution anament. Sewage work, J. 19 ; 629–644.

Waahaman, B.O. Harris, D.O. and Willium, B.C. (1970). Flood of December 1904 s January 1965 in the Western State. 11 Stream flood and sediment data wat supply pop. VS. Geal. Surv. 1866.

Webb, L.J. (1985): "An investigation in to the occurrence of sewage fungus in Rivers containing Paper mill effluents". Wat. Res. Vol. 19 No. 8 : pp. 955–959.

Webber, W.J. and Stumn, W. (1963). Mechanism of hydrogen ion buffering in natural waters, J. Amer. Wat. Works. Assoc., 9 : 1553.

Weissman, S. and Kott. Y. (1979). The effect of changing oxygen concentration on the viability of bacteria, P. Symp. Isreal Ecol. Soc., A–144.

Welch, P.S. (1951). Limnology. McGraw Hill Book Co., New York.

Wnipple, G.C. and Parker, H.N. (1962). On the amount of oxygen an carbonic acid dissolved in natural waters and effect of these gases upon the occurrence of microscopic organisms. Tran. Amer. Micro. Soc. 23: 103–144.

Wurtz, C.B. (1955): "Stream biota and stream pollution". Sew. & Inds. wastes vol. 27: pp. 1270–1278.

Yadav, Y.S., Singh, R.K., Choudhury, M. and Kolekar, V. (1987). Limnology and productivity of Dighali Bheel (Assam) India. Trop. Ecol. 28(2): 137–147.

Zafar, A.R. (1964). On the ecology of algae in certain fish pond of Hyderabad India I. Physico–chemical complex Hydrobiologia 23: 179–195.

Zahid, Waleed A. M. and Ganczarezyk Jerzy J. (1990). Suspended solids in biological filter effluents, Wat. Res. Vol. 24(2): 215–220.